AF556735

CAGE AND AVIARY SERIES

THE WORLD OF ZEBRA FINCHES

OTHER BOOKS AVAILABLE

Gouldian Finches
A. J. Mobbs

Cult of the Budgerigar
W. Watmough

The Budgerigar Book
Ernest Howson

Exhibition Canaries
Joe Bracegirdle

The Lizard Canary and Other Rare Breeds
G. T. Dodwell

The Yorkshire Canary
Ernest Howson

The Roller Canary
George Preskey

Shackleton's Yorkshire Canary

The Pheasants of the World
Dr. Jean Delacour

Hummingbirds
A. J. Mobbs

THE WORLD OF ZEBRA FINCHES

by
CYRIL H. ROGERS, F.B.S.A.

Colour Plates
by Frances Fry

Joint Publishers:

NIMROD PRESS LTD
Liss, Hants, GU33 7PR
England

SILVIO MATTACCHIONE & CO.
Pickering, Ontario
Canada, L1W 3H2

ISBN – UK – 1-85259-012-2
ISBN – Canada – 0-9692640-2-X

First Edition 1986

Restricted Distribution

Books are sold on the understanding that there will be no transferring of copies between the two distribution areas by purchasers in the respective areas. It is therefore illegal for a bookseller *not* resident in North America to sell in the U.S.A. or Canada and similarly sales should not be made from North America to the rest of the world.

Joint Publishers

NIMROD PRESS LTD
Liss, Hants
England, GU33 7PR

World distributors excluding North America

SILVIO MATTACCHIONE & CO.
1020 Brock Road
Pickering, Ontario
Canada, L1W 3H2
Distribution North America, (Canada and U.S.A.)

CONTENTS

MONOCHROME ILLUSTRATIONS

COLOURED PLATES

All the known mutations have been painted by artist Frances Fry. She has been guided by the author and Michael S. Wrenn, a prominent breeder and member of the Zebra Finch Society. They appear after page XVI.

ACKNOWLEDGMENTS

My thanks to individuals and companies who assisted and/or gave permission to reproduce. These are:

Zebra Finch Society, Great Britain
Zebra Finch Society of Australia
Cage and Aviary Birds, London, Great Britain
C. af Enehjelm, Denmark
Nils A. Svensson, Sweden
M. Wrenn, Birmingham, Great Britain
J.A.W. Prior, London, Great Britain
W.L. Cotta, California, U.S.A.
Blandford Press, Poole, Great Britain
Serventy & Whittell, Australia

CYRIL H. ROGERS

PREFACE

Some fifty years or so ago Zebra Finches were just hardy little Foreign Birds ideal for a mixed aviary of small exotic birds. As they were found to be excellent breeders and very useful as foster parents to the more rare foreign species this increased their popularity with bird breeders. A few colour mutations then began to arrive and this helped to make them more widely kept, bred and exhibited.

In 1952 a small group of very keen Zebra Finch breeders met in Birmingham and founded the **Zebra Finch Society** which today is the largest specialist society catering for a one-time foreign finch-like bird. Then in 1958 this progressive society declared that **Zebra Finches** should become a fully domesticated species and no longer to be classed as Foreign Birds and this decision was accepted world wide. As more colour mutations arrived on the scene they greatly helped to increase further the number of breeders in most countries.

At all bird exhibitions, both local and open they have a special section for Zebra Finches with the Zebra Finch Society acting as a guiding hand in all matters appertaining to these delightful little birds. Although there are a number of exhibitors of Zebra Finches all over the country, there are most certainly an equal number who maintain aviaries of mixed-coloured Zebra Finches just for pleasure and decoration.

The object of this book is to help in making their culture even more widespread and to give the history of their development.

Cyril H. Rogers
Aldeburgh, Suffolk
June 1986

AUSTRALIA – DISTRIBUTION OF THE ZEBRA FINCH IN THE WILD

Zebra Finches are now considered to be fully domesticated birds. However, they still exist in the wild and it is useful to see their natural habitat which gives a better idea of how they live.

As shown on the distribution map opposite, zebra finches are to be found in many areas throughout Australia, but they are found in abundance in the dry interior of the Continent.

Zebra Finches are the most common species of grass finch to be found. Wherever water is provided, whether naturally or artificially, they exist, nesting in bushes and small trees. Apparently they also nest on the ground or in rabbit burrows. The nest is made of a mixture of grasses and lined with feathers, fur, wool or similar material.

In the wild they lay 4 to 5 eggs, but this may vary plus or minus 2. Incubation varies from around 12½ days to 16 days depending upon conditions (11-12 days in captivity). In the wild breeding depends upon rainfall, but in captivity there are no breeding problems so a number of broods may be raised in a season.

For further details the reader is referred to *Australian Finches*, Klaus Immelmann, Angus and Robertson, Sydney. This book contains many interesting facts on Australian Finches from observations and studies made by the author in Australia.

DISTRIBUTION MAP

Breeding Area for Zebra Finches
(covers most of Australia)

The Ideal Shape for all colours. An early painting by R. A. Vowles, the famous cage-bird artist. (Original in colour)

DEDICATION

I dedicate this book to my dear wife Helen
who was so helpful for very many years
and who was my delightful partner
in writing my many books on
cage and aviary birds

A Florida Fancy White Cock (foreground)

COLOUR IDENTIFICATION PLATES

The Plates which follow have been painted by Miss Frances Fry, the well known bird artist. They portray the colours or mutations now available showing the extent to which the enthusiasts have gone in breeding the many varieties now available.

In practice many variations will be found in the colours of the same mutations, but bred in different places and by different breeders. The paintings should be used as a *guide* to what might be expected. Some of the colours (as indicated in the text) are quite rare and very few people have seen them.

The Author, Cyril H. Rogers (with notepad) judging a club show

Plate 1. Zebra Finches Normal (Grey)

Plate 2. Zebra Finches. **Top Left** Dark Fawn. **Top Right** Pied Fawn
Lower Left Light Fawn. **Lower Right** Cream.

Plate 3. Zebra Finches. **Top Left** Silver. **Mid Right** Marble. **Lower Left** Grey (or lead) cheek lobed. **Lower Right** White.

Plate 4. Zebra Finches. **Top Right** Pied Grey. **Mid Left & Centre** Pied Fawn. **Lower Left** Penguin Grey. **Lower Right** Penguin Fawn

Plate 5. Zebra Finches. **Top Left** Yellow Beak Grey. **Top Right** Yellow Beak Fawn. **Lower Left** Crested Penguin Fawn. **Lower Right** Crested Grey.

Plate 6. Zebra Finches. **Top Left** Florida Fancy Cream. **Top Right** Florida Fancy White. **Mid Right** Saddleback. **Lower Left** Chestnut Flanked White.

Plate 7. Zebra Finches. **Top Left** Light (or bright) back. **Mid Right** Light red cheek lobed. **Lower Left** Orange breasted. **Lower Right** Grizzled.

Plate 8. Zebra Finches. **Top Left** Melanistic. **Mid Right** Black. **Mid Left** Black Breasted. **Lower Right** Schwarzling.

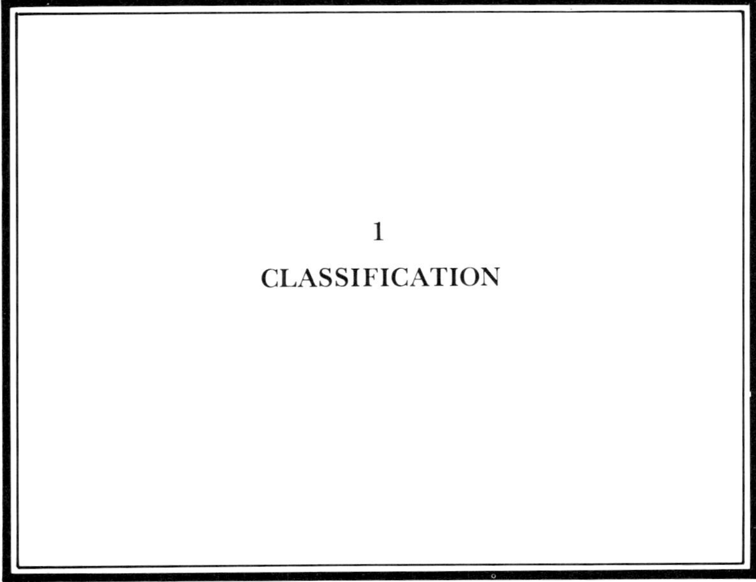

1 CLASSIFICATION

"Inhabits Northern, Western, and Southern Australia, whence it is regularly and abundantly imported; it is hardy, easily bred, and very beautiful"

Dr. Arthur G. Butler
A Victorian Naturalist and Writer

A lead-cheeked Cock

Chapter 1

CLASSIFICATION

There have been several different names given to the species of birds now universally known as Zebra Finches. After searching through the literature appertaining to early ornitholigical field work in the Australian area, I have been unable to find just when Zebra Finches were discovered. In 1840 the celebrated Naturalist John Gould sent the first Budgerigars to Great Britain and two years earlier he had given a description of the species now known as Zebra Finches and it is thought that amongst other Australian birds exported to Europe during that period there could have been some Zebra Finches. Later in this Chapter readers will find reference to the data of other European Naturalists.

Zebra Finch Names

The name used in Britain is Zebra Finch and the Latin names are:

Taeniopygia castanotis
Taeniopygia guttata
Taeniopygia guttata guttata
Taeniopygia guttata castanotes

In the 1970s the Zebra Finch was placed in the same genus as the Grass Finches and the old prefix was replaced by *Poephila*; thus:

Poephila guttata guttata
Poephila guttata castanolis

Names used in other countries

Australia	Zebra Finch, Chestnut-eared Finch, Pimp.
France	Moineau mandarin, Diamant mandarin.
Germany	Zebrafinken.
Denmark	Zebrafinke.
Holland	Zebravink.
Sweden	Zebrafinker.
U.S.A.	Zebra Finch.

Native Aboriginal Names:

Chiaga, Ymi-eye, Neamoora, Newmerri, Nyi-Nyi and Nye-Nye.

There are of course many other Aboriginal names for this widely distributed species according to the area in which the tribes live. It will be noted how much these names differ and that the last two resemble the call note of the Zebra Finch.

HABITAT

Zebra Finches are to be found in all the States of Australia and some of the neighbouring Islands in varying densities where there is water, suitable breeding and feeding areas. They like to nest in the thickets of various native shrubs and bushes and feeding on the many grasses and weed seeds; also taking any small insects or grubs they come across in their search. They make small rough-domes nests of both stiff and soft grasses, rootlets, mosses, lining with vegetable down and small soft feathers or wisps of sheeps wool. The nests are often built in communities of between ten to fifteen nests in a large bush or a clump of bushes. Clutches of white eggs range from three to eight with four or five being the average which is similar to the clutches as laid by the domesticated strains. The average size of the wild laid eggs is around 15.0 x 10.5mm and those of the domesticated races are often slightly larger. Both sexes share in the incubation which lasts some eleven to twelve days. As the chicks develop it is the cock birds that seem to do the largest share of foraging for food. Young birds will leave the nest when twenty-one to twenty-four days old and for a time are looked after by the parents, mainly the cock birds. A similar pattern will be found with the domesticated strains.

DESCRIPTION

Adult Cock Eyes dark. Beak coral red with narrow black stripe at sides followed by a wider white one and then another black stripe (tear mark). Cheek lobes deep reddish orange. Side flankings rich reddish brown well sprinkled with small round white spots. Head dark grey with some darker ticking on top. Neck, back and wings grey. Chin, throat and upper breast white with fine black zebra markings ending at base with black bar about 1/8th" wide. Under parts to vent white. Rump white with black at sides. Tail black heavily barred with white. Feet and legs reddish brown.

Adult Hen. Similar to cock but minus cheek lobes, side flankings, throat and chest markings and pale grey underparts. Beak paler red.

Juvenile similar to hen but with black beak.

Sub-Species

Various Ornithological writers give different numbers of sub-species that are said to exist in the widely separated areas of Australia with at least one coming from the Islands of Timor and Flores. In their excellent book *Birds of West Australia*, published in 1962, Serventy and Whettell give eleven sub-species which differ somewhat in size and depth and distribution of markings and colour. The late G.M. Matthews gives seven sub-species in his book *Systema Avium Australsianarum.* Neville W. Cayley, F.R.Z.S. in his book *Australian Finches in Bush and Aviary* also gives the sub-species as seven and some other writers quote the number from four to eight. From these different figures it can safely be said that a number of sub-species can be identified in the field. As there is such a variation in their actual number it may well be that some have a very localised habitat and can easily be missed when a survey is being made. It is known that mutant colours of Zebra Finches do occur in the wild and in the past such birds may have been included amongst the sub-species by some writers before they were recognised as colour mutations. One sub-species that is definite is the race that inhabits the Timor and Flores Islands and here the throat markings differ from the normal.

In a most instructive article that appeared in the 1961 Zebra Finch Society's Year Book my friend Curt af Enehjelm, one time Curator of Helsinki Zoological Gardens, Finland, gave some enlightening details of the early writings on Zebra Finches. This article was based on his book *Zebrafinker Magefinker og resfugel*, published in 1959.

Below I give some extracts from this article:

> **. . . In 1838 John Gould gave a description of the Zebra Finch. The bird was, however, described earlier. The French Naturalist L.J.P. Viellot (1748–1831) in his wonderful book** *Histoere naturells des plus beaux oiseaux chanteurs de zone torride* **(1805) gave a rather correct description of the bird (Zebra Finch) which he called Le Bengali mouchete, however with the wrong statement that it originates from the Moluccan Islands. This is probably the first description of the species, but it is a question, if the bird described did not belong to the Island sub species from Timor and Elores than from the Australian mainland.**
>
> **It is more than probable that Mon. Viellot himself never possessed living specimens and in any case this does not appear from his report. He only states that he got the bird through a certain Mon. Becaeur but not a living bird, a skin or only a description . . .**
>
> **There are no statements when or by whom the Zebra Finch was first brought alive to Europe, and any account about the first breeding in captivity. Certainly Dr. Karl Russ, states in some editions of his books, that the bird was already bred by Viellot . . . In the first Edition of his** *Prachtinken*, **he writes: Viellot has bred this bird (Zebra Finch) and also gives a picture of the juvenile**

plumage.

As mentioned earlier, nothing in Viellot's own account confirms that he had bred the bird or even had living specimens . . . It will therefore hardly ever be told when or by whom the first Zebra Finch was first bred in captivity. No doubt, however, this actually happened at a comparatively early date. In the first volume of the German paper *Die Gefiederte Welt* of 1872 the breeding of Zebra Finches was already considered quite common and in the first Edition of his earlier book *Die Prachtfinken* Dr. Karl Russ writes 'that the experiences from the breeding of Zebra Finches are fundamental for all breeding of cage birds'. . . In a number of books written in the latter eighteen hundreds and early nineteen hundreds in the English language the Zebra Finch and its treatment in captivity is elaborately described. *Foreign Cage Birds* by C. W. Gedney, *Birds I Have Kept* by W. T. Green, *Foreign Birds* and *Foreign Birds in Captivity* by A. G. Butler are amongst the most interesting volumes.

Mutations

Mutations had obviously appeared in the time of Dr. Russ's third Edition of his book *Fremdlanische Stubenvogel* he writes that 'Zebra Finches have already been bred in different colours, but these are not of any importance. It is very hard to understand why Dr. Russ who was so very interested in yellow and blue Budgerigars took such an unsympathetic attitude to the new colours in Zebra Finches, and that he did not give a description of them . . .

OTHER WRITERS

The late Allen Silver F.Z.S. who was a famous Naturalist writer, judge, exhibitor and breeder of many species of both British and Foreign birds told me that the first book in English on Foreign birds was the one by C. W. Gedney. In this book there is a full account of the Zebra Finch and its behaviour, and also its breeding in captivity.

From the above extracts it can be gathered that Zebra Finches were freely breeding as cage and aviary birds in Europe and Great Britain in the late eighteen seventies. It is therefore a pity that Dr. Russ did not think it necessary to record descriptions of the colour mutations that he had seen during that time. They may well have been the darker shades of colour changes, for if they had been White or Chestnut-flanked White, Dr. Russ could not have failed to have recorded them. Unfortunately these early mutations are things that must remain a closed book for ever.

Zebra Finches Nesting (Chestnut Flanked Whites, Normal Grey Pieds and Youngster)

2
FIRST MUTATIONS

AN EARLY DESCRIPTION

The male bird is barely 3in. long, of a robust shape, and closely feathered, his prevailing body colour above being delicate pearl grey and white beneath, primaries of wings and tail black, the tail being barred with white stripes. The beak and legs are orange pink, the head and neck are finely marked with zebra-like stripes of black, terminating in a black band upon the breast, the abdomen being pure white. The cheeks are marked with a circular patch of chestnut brown (this led to the bird being called the "chestnut eared finch", by Gould), and the eyes have a reddish tinge in them. The hen is slightly smaller than her mate, and although of the same grey body colour, yet she lacks his zebra markings and also the brown cheek patches, the absence of which is a sure indication of the sex that purchasers should remember.

C. W. Gedney *Foreign Cage Birds*

THE ZEBRA FINCH.

(*Amadina Castonatis.*)

From *Foreign Cage Birds*
(G.W. Gedney)

Chapter 2

FIRST MUTATIONS

It appears from the records that the first mutation to be fully established as breeding strains were the Whites in 1921.

WHITES

The first Whites to be recorded appeared in an aviary containing a mixed collection belonging to A.J.Woods of Sydney, New South Wales, who established the mutation. It was quickly discovered that these White birds were of the Recessive breeding kind and consequently it took a few years to build up a reasonable stock. The records do not state whether the first white birds were pure in colour or if they carried any grey flecking on their backs. At the present time both pure and flecked Whites are bred, the former mostly by careful selected pairings and the latter mostly from uncontrolled collections. In due course of time examples of the Australian Whites were exported to various countries including Europe and Great Britain.

Our present day Whites appear to have all descended from this Australian mutation although of course it is always possible that further mutations of the same genetical kind may have occurred. As their colour, or rather lack of colour, is so different from the Normal Grey the sudden appearance of a White in a mixed breeding colony would at once be spotted. With some of the darker coloured mutations it would have been quite easy for them to have been overlooked by the untrained eye, but certainly not so with a White. It was not until about the early nineteen-hundreds that the first Whites were bred in Great Britain, but I can find no trace of a record of the first actual breeders.* As far as the records go their first appearance on the show bench at a large National show was at the 1936 Crystal Palace Exhibition where C. Hooker exhibited a single White, sex not stated, although the Normal Greys have been exhibited in small numbers for a considerable time. It was not until 1948 that White Zebra Finches were seen again at the "Crystal Palace" Championship Show held at the Royal Horticultural Hall, London, the old Palace having been burned down in 1935.

During my various discussions with the late Allen Silver whose memory of ornithological matters stretched back into the eighteen-nineties I learned that normal Grey birds with odd white feathers in tail, wings and on the body, were to be found in the importations of Continentally-bred Zebra Finches. In those days such specimens were always discarded as being mismarked birds and no attempt to form breeding strains from them was taken. It is quite possible that if planned pairings were made Pied birds could then have been evolved.

At the 1937 Crystal Palace Exhibition two pairs of White Zebra Finches were benched by T. Crewes and H. J. Indge and the former also exhibited a pair of Fawns and a pair of Normal Greys. During this period White and other coloured Zebra Finches were usually exhibited in with Foreign Finch Hybrids and abnormally coloured Foreign Birds and this was so until the nineteen-forties. From this time onwards Whites in varying numbers had been shown at the majority of exhibitions where classes for Zebra Finches only were scheduled.

Undoubtedly the highlight of the White mutation and for all Zebra Finches for that matter was when L. Harris of Birmingham showed a pair of Whites bred by himself at the 1972 Exhibition of Cage and Aviary Birds at Alexandra Palace, London, and won the Supreme Award for Best Birds in Show. As one of the four Judges who selected this pair I know first hand what delightfully pure coloured, shapely and in immaculate condition these birds were and certainly deserved their tremendous success.

Early on during the development of Zebra Finches into a

In 1933 at Keston Foreign Bird Farm it was reported that a practically pure white hen had been bred.** ***Avicultural Magazine**, 1934, p.53.**

fully domesticated species Whites were always paired to Whites without any special selection for purity of colour. When Zebra Finches really got going as exhibition birds, breeders of Whites started to select their breeding pairs for freedom of flecking and by doing this they met with considerable success. The number of pure Whites produced increased substantially, but unfortunately substance and type did not follow in the same pattern. After considerable thought some breeders decided to cross their best Whites to the bolder more typey Fawns and then cross the young Normal Grey/Fawn White cocks and Fawn/White hens together. It must be remembered that at this period all Whites were masking Normal Grey so both cock and hen Fawns were used as outcrosses. In a very short time there was a distinct and noticeable overall improvement in the Whites bred that were masking Fawn. A little later on another school of thought used Pieds to cross into their White strains and again the plan proved highly successful in most cases. At the present time Whites can mask many colours although the best ones seem to emanate from the early introduction of the Fawn and the Pied characters.

The Zebra Finch Society Standard for exhibition of White Zebra Finches is –

COCK AND HEN: Eyes dark. Beak coral red. Feet and legs pink. Plumage pure white all over. Hens usually have beaks of a paler shade of red. Flecking of any colour on back, body, wings or tail are show faults. Bodily fitness together with perfect condition of all feathers is essential.

FAWN

The next colour mutation to be recorded was the Fawn and the first birds to appear in the wild in their native Australia during 1927. At one time it was thought that Fawns came into being some time after World War II and originated in South African aviaries. However, this assumption was incorrect as the following extract from the August Issue 1958 of the magazine of the Avicultural Society of Australia proves.

. . . I will make a complete coverage for the benefit of all who are interested and may have been misinformed. Many are of the opinion that these birds (Fawns) made their first appearance after the second World War. This is not so as, the first were bred as Fawns and established in 1927 and, for a number of years, were confined to Australian aviaries Then a few years before the War a few pairs went to a fancier in South Africa which were fully established Fawn mutation of the Zebra Finch. . . . Prior to the export ban being imposed here in Australia, many Fawn Zebra Finches which consisted of surplus stock were sent to fanciers overseas and aviculturists throughout the world were able to obtain and specialise and consequently establishing the species.

However the following information should serve to clarify the whole situation. This I feel would also be interesting news to all Zebra Finch enthusiasts throughout the world, particularly as it started a new area in Aviculture. This phase of Aviculture may have been lost if one of our local Fanciers had failed to be observant in detecting the original birds in the wild and if breeding attempts made by his brother had failed. The credit for our good fortune goes to Mick Lewitzka, who, at the time the birds appeared, was in residence of Coward Springs (five hundred miles north of Adelaide) and his brother, Fred Lewitzka, who pioneered the groundwork in establishing a strain which later accumulated in the production of the Fawn Zebra Finch – the initial introduction of which took place in 1927.

The original birds were observed in a locality well outside the general rainfall area where there is a lack of vegetation so abundant in the higher rainfall areas . . . However, at night, temperatures usually drop considerably and, consequently early morning temperatures are also very low. The camp fire is therefore a prominent feature in the life of the outback residents and, around these, centres the initial story of the origin of the Fawn Zebra.

The original birds were two sports of a fawn mutation, and were observed flying around with a large flock of Normal Grey Zebras. In seeking warmth these birds with their normal coloured grey mates, would venture close to the hot coals and ashes of the dying fires on cold mornings. Many hundreds of these birds were trapped over and over again, until the two sports were finally caught. The two birds were hens and both of the Fawn mutation, carrying the markings of the true Fawn of today. Having been trapped, they were promptly despatched to Adelaide and the slow and tedious task of breeding progeny similar to the two trapped birds was begun.

For several years, all the abnormal birds bred were hens and were not Fawns. They were a poor blotchy silver and, when the first males were bred, they carried very little colour in their ear patches. Being of this colour these birds in many instances were referred to as 'Smokies'. Upholding their reputation these birds soon produced quite a little colony and their numbers were a little high for Fred Lewitzka to handle, and some were farmed out to fanciers who made good use of them.

It will be seen that it took the original breeder a number of generations to produce Fawn cocks not having the knowledge of bird genetics as known at the present time. At first only hens were bred and this would be expected from the use of young cocks bred from the two wild caught Fawn hens. Now comes what I think is the rather puzzling part of the extract where it states " . . . *all the abnormal birds bred were hens and they were not Fawns. They were a poor blotchy silver and when the first males were bred they carried little colour in their ear patches.*" This seems to indicate that either the two original Fawn hens were definitely Fawns or the Greys they were paired to may have carried unknowingly a Recessive Dilute character This could account for the blotchy silver colour and the pale ear patches of the first cocks. It is fairly certain that selective breeding took place to reproduce the fawn colour and consequently bred out the blotchy greyish Fawns in the first place. I do not think however that we shall ever get the real answer as these events happened over fifty-five years ago, but we do have the lovely Fawns as a lasting result of the Lewitzka brothers endeavours.

The above records indicate quite clearly where and when the first Fawn mutation originated and was developed. I have not been able to trace just when the first specimens came to Great Britain but it appears that the first Fawns exhibited at a National Exhibition was in January 1948 when a pair was benched by T. Crewes of Beckenham, Kent, at the Royal Horticultural Hall, Westminster, London. From this it would appear the Fawns came to Europe and Great Britain very soon after the end of World War II as I can find no trace of their presence in 1939 or 1940. Just prior to the formation of The Zebra Finch Society in 1952 Zebra Finches had started to become desirable exhibition birds and in fact the first separate class for New Coloured Zebra Finches was scheduled and nineteen entries were received which included Whites, Pieds, Fawns, Silvers and 'Blues'.

It is interesting to note that in this New Coloured Class were a pair of Fawns and a pair of Cinnamons entered by the same exhibitor. This indicates that the knowledge about New Colours possessed by the then Zebra Finch breeders was very limited. Within a short period however breeders came to recognise the fact that the lighter Fawns and the darker Cinnamons were actually due to the same mutation. As there were light and dark coloured Normals so this was reflected in the different shades of fawn.

The first National Exhibition to put on a class for Fawns was held at Olympia, London, in January 1954 and was for Fawns and Cinnamons and attracted some twenty-one exhibits. In the following year this class heading was altered to read Fawns, Light and Dark, as by then it was realised that Fawns and Cinnamons were one and the same mutation. Amongst the exhibitors in this class were the late Peter Pope, one time Chairman of the Z.F.S., the late Stan Moulson, Hon. Secretary Z.F.S. from 1952-62, Raymond Sawyer the well known Judge and exhibitor of Foreign Birds, and Ted Hounslow who was Ring Secretary to the Z.F.S. for many years. Fawns have continued to be one of the most popular mutations with breeders and exhibitors from this period up to the present time and at most shows they are still the largest Section. One of the most successful exhibitors of Fawns at the National and other larger Exhibitions is J. E. Addison of Carlisle whose birds have won Best in Show at the National Exhibition no less than sixteen times to date. Being a sex-linked colour hens are more plentiful than cocks and it is far easier to find good exhibition hen birds that it is cocks. This is true with all the sex-linked varieties and can make the matching of exhibition pairs a little more difficult than with the Recessive and Dominant mutations.

The Zebra Finch Society's *Standard* for exhibition Fawn Zebra Finches is:

COCKS: beak red, feet and legs pink, head, neck and wings deep even fawn. Breast bar dark. Throat and upper breast light fawn, with zebra lines running from cheek to cheek continuing down to breast bar. Underparts white, may have some fawnish shading near vent and thighs. Cheek lobes dark orange. Tear markings the same shade as breast bar. Tail dark barred with white. Side flankings reddish brown with even clear white spots.

HENS: as other hens but of the same shade as fawns as cocks. Fawns can be of a number of different depths of colour and it is important to have both members of exhibition pairs of the same depth of colour.

CHESTNUT FLANKED WHITE

It would seem that the next mutation to appear amongst Zebra Finches was the Chestnut-flanked White and like the Fawn the original stock were taken from a wild flock in Queensland, Australia, some time during 1937. This second sex-linked mutation was more quickly developed than the Fawn but even so I do not think they became known in Europe or Great Britain until after World War II. At first their Australian breeders called them Marked Whites as of course they were. After due consideration it was decided that the term 'marked White' equally applied to heavily flecked specimens of the original Whites and the birds called Saddle-backed Whites so the name Chestnut-flanked White was univerally adopted by Zebra Finch fanciers. The name 'Marked White' is sometimes used by European breeders, this is due to the fact that young birds in nest feather show a lot of blackish shading on their face and head. As this is only descriptive of immature birds it is not a good name for the mutation. Another name used by European breeders to describe Chestnut-flanked Whites is 'Marmosette' and will often be found in their Avicultural literature.

Some of the first specimens to come to Great Britain were I think imported from Europe by the Keston Foreign Bird Farm and those I saw in the early nineteen-fifties were near white in colour with very heavy markings but rather leggy in type. Although Chestnut-flanked Whites had been exhibited in small numbers at the National Shows during the middle fifties it was not until 1974 that they were given a separate class at this great Show. The class attracted ten entries and was won by a good coloured pair shown by G. Douglas of Annan, Scotland. It was however in 1957 that The Zebra Finch Society recognised the name Chestnut-flanked White for this mutation and evolved a show standard and began providing separate classes for .this birds at their Patronage Shows.

Amongst the early breeders and exhibitors of Chestnut-flanked Whites C. H. Smith, S. W. Moulson, G. B. Shaw, A. J. Rooke, Peter Pope, Mrs. E. L. Ellis, W. J. C. Hocking, D. Norris, T. Foster, were those I can clearly recall having judged birds at various shows. The present-day Chestnut-flanked White enthusiasts and keen exhibitors include J. R. Addison, T. A. Ottaway, I. E. Whiston, Mr & Mrs Binns, D. Yates, S. Draycott, D.Gould and D.W.King, together with many other interested Fanciers and the mutation is steadily expanding.

From the time Chestnut-flanked Whites were first imported into Great Britain from Europe and some time after 1946, most likely between 1956 and 1970, a further mutation must have occurred which had a much more cream coloured body shade than the original birds. These creamy coloured birds had paler markings that the whiter ones but they had more substance and generally better type. For these reasons they made a bigger impact at the shows and their popularity grew especially as both shades could be paired together giving *all* Chestnut-flanked White young showing they belonged to a similar mutation. In the November 1970 the Zebra Finch Society Newsletter amongst the Editor's notes was some information from Australia from which I quote:

> **. . . Mr Thornley also gives some further interesting information about Zebra Finches in Australia, it appears that they have two kinds of Chestnut-flanked Whites, one with normal markings and one with dilute markings. As breeders know we have a variation in depth in the markings of our own birds ranging from quite deep to very pale, and I am wondering whether the Australian birds fall into the same category . . .**

In more recent years and especially so since the Light-back mutation was imported from the Continent heavily marked birds have appeared again and have been called Continental Chestnut-flanked Whites. These are of course from the same mutation that was originally imported and coincide more closely with the Zebra Finch Society's description of exhibition type Chestnut-flanked Whites. Much has been written about these birds in the Fancy Press and the creamy coloured Chestnut-flanked Whites and their relationship. It appears that there is a difference in the skin colour of the newly hatched chicks one being darker than the other. The ones with the dark coloured skins are the more recently imported European kind. Like the Opaline variety in Budgerigars, where some four different mutations of similar pattern have been crossed and from which a *Standard* has been formed, the two and probably more mutations of the Chestnut-flanked White will most likely follow the same course.

The Chestnut-flanked Whites are the second sex-linked variety to have their origin amongst the wild Australian flocks and being of this particular method of reproduction it is naturally far easier to produce hen birds than cocks. For this reason it is easier to breed first rate quality exhibition type birds than it is cocks. When these two different sex-linked mutations are paired together one behaves as though it were a Normal and this of course is what would be expected. For instance, if a Fawn cock is paired to a Chestnut-flanked White hen the resulting young will be Normal/Chestnut-flanked White Fawn cocks and Fawn hens. Reverse the mating and the young would be Normal/Chestnut-flanked White fawn cocks, like the previous pairing, but the hens would be Chestnut-flanked Whites. A Normal/Chestnut-flanked White Fawn cock paired to a Chestnut-flanked White hen would give two kinds of Chestnut-flanked hens amongst their young, one masking Normal and the other masking Fawn, although both will have the same coloured maskings. The same colourings would be found in cock birds of the same genetical composition.

Although the change of colour does not affect the overall pattern of the Chestnut-flanked Whites, certain varieties can do just that and some rather interesting composite birds have been bred. From an exhibition angle the resulting young may not be so attractive as their straight ancestors but colour breeding enthusiasts find them most intriguing. One of the first examples of such a combination I saw was at an Exhibition during the late 1960s. In this case it was a Recessive Pied and Chestnut-flanked White and it was interesting to see the broken pied markings on the near white ground colour blotched with pure white. I have seen a number of like examples since and they have all shown clearly their mixed parentage.

A few Penguin Chestnut-flanked Whites have been bred and the only one I have seen was a cock bird which I thought rather insipid in its colouring. This of course would be expected with both parents being Dilutes. Although quite a curiosity it is not a composite form to be encouraged. On the other hand the Black-breasted Chestnut-flanked Whites are differently marked, showing strongly the Black-breasted pattern. At the moment only an odd specimen or two have been bred in this country but I think Continental breeders have been producing them fairly regularly.

There is a variance of opinion as to the best outcross for maintaining and improving the overall exhibition qualities of Chestnut-flanked Whites, some say it can only be done by selecting within a Chestnut-flanked strain and others that good exhibition type Normal Greys should be used for the purpose. In my opinion, based on a number of years observing different strains, the most satisfactory results can be achieved by the combining of both methods. In all cases the Normal Greys used should be pure bred, well marked and free from brownish or pale markings on wings and very clear deeply coloured flankings, cheek lobes, breast and tail barrings. It will be noticed that Normal Greys of different strains vary quite considerably in their colouring and the selection of the right coloured birds as mentioned above is most essential in developing Chestnut-flanked Whites. At the same time the Chestnut-flanked Whites used should be selected for their depth of markings and purity of colour. If this method is practised for several seasons a very marked improvement in the overall colour of the Chestnut-flanked Whites will be achieved. Type and substance should not be neglected whilst these pairings are being made in building up an exhibition strain.

The Zebra Finch Society's *Standard* for Exhibition Chestnut-flanked Whites is:

COCKS: eyes dark, beak (coral) red, feet and legs (deep) pink. Head, neck, back and wings as white as possible. Underparts pure white. Breast bar as near black as possible. Tear markings same shade as breast bar. Cheek lobes orange. Tail, white with bars, colour to match breast bar. Flank markings brown with clear white (round) spots.

HEN: As other hens to match cocks. May have light head markings

DILUTES – SILVERS AND CREAMS

It is usual when a species becomes thoroughly domesticated

that dilutations of their colour occurs and in 1932/3 this happened to the Zebra Finch. Once again the original examples of this mutation were bred in their native Continent of Australia and the first arrivals came to Europe in about 1938. Silvers as these birds were called were soon being bred in Continental aviaries and at the Foreign Bird Farm, Keston, England. The colour of these birds was about half that carried by Normal Greys and like them could and did vary in their actual shade. From the onset it was realised that the Dilute Silvers followed a Dominant breeding behaviour and could be had in two genetical kinds, Silvers having only a single quantity of the Dominant Dilute character paired to Normal Greys gave 50% of each colour. Birds having a double quantity when paired to Normal Greys gave all Silver young with a single quantity of the character. When two single quantity Silvers were mated together they gave 25% Normal Greys, 50% Silvers single quantity and 25% Silvers double quantity which of course is in line with the expectation of a Dominant colour character. It must be pointed out here that although these Greys had two Silver parents they were still just ordinary Greys.

It was not long before British breeders raised Dominant Creams, the Fawn Dilute, which had already been produced in Australia. In fact it has been said by some writers that these Creams actually appeared in Australia before the Silvers. Silvers do carry quite a heavy cream shade on their backs which of course is undesirable. I have not yet been able to verify which came first as in these early days of development of Zebra Finches it has been very difficult to ascertain accurate details. No matter which came first both shades were very welcome additions to the growing list of Zebra Finch colours.

Amongst the early breeders and exhibitors of Dominant Silvers were T.R.W. Crewes, Stan Moulson, Peter Pope, H. Bernard Smith, G.B. Shaw, J. Crozier, W.J.C. Hocking, C.E. Stathers, T.Foster and others.

After the first burst of enthusiasm for the new colour form their popularity began to wane as did their overall purity of colour. As these Silvers lost ground both as aviary and exhibition birds so the Dominant Creams (Dilute Fawns) took their place. Over the past years some very beautiful pairs of Dominant Creams have been exhibited up and down the country frequently gaining Best Zebra Finches in Show both as adults and young birds. The first time I can trace where Dominant Silvers were exhibited was at the January 1948 Crystal Palace Show held at Royal Horticultural Hall, Westminster, where a pair were shown by T.R.W. Crewes. However, they did not appear again at a National Show until the National Exhibition of Cage Birds at Olympia, London, November 1952, where pairs were shown by Peter Pope, Mrs. P.E. Alner, S. Moulson and S.N. Pinfield. In the same class R.C.J. Sawyer benched a pair which he called 'Blue', but I cannot recall their being any difference in their colour to the Silvers. I will have more to say about the 'Blue' phase later.

RECESSIVE DILUTES

Now to return to the Dilutes as in addition to the Dominant breeding kind there also exists several races that are Recessives. The origin of these Recessive Dilutes is not known for certain but it does seem they occurred at approximately the same time as the Dominant kind. The first to be noted were the Recessive Silvers or as they were called at the time Blues. Although these birds were more of a bright bluish grey than the Dominant Silvers they were certainly not the blue of the Budgerigar and are best known as Recessive Silvers. Before the Zebra Finch Society was launched Silvers were mentioned in various Ornithological publications although at first no clue was given as to their breeding behaviour. It is therefore not certain that when Silvers were mentioned whether they were Dominant or Recessive as in the early days precise details of colour and genetics were often vague.

One of the first breeders of Silvers in Great Britain whose original birds appeared as a mutation was C.E. Stathers of Hull, Yorkshire. I quote from an article he wrote in The Zebra Finch Society's *Year Book and Bulletin* of 1958:

In the 1957 Year Book I was interested to read the articles by our

various members and in particular C.af Eneljelm of Finland re Blue or Recessive Silvers. I have bred some and I called them slate coloured at first, but have shown them a few times as Silvers ..

In September 1953, I purchased a pair of Zebra Finches from a local Fancier. The cock had white flights and a white chin and a small white patch on the head, its parents were a Normal cock and a White hen, the hen was a normal, its parentage I cannot say.

I put them in a cage in the kitchen recess and in February 1954 I put a nest box in after two lots of clear eggs, I got my first young ones, a nicely marked Pied cock and two hens and I finished the season with twelve young ones which included the Pied cock and twelve Normal cocks and hens. I built a small aviary, and parted with all but two pairs of the young ones, and a pair of unrelated birds. I mated the cock to one of my young hens and hen to a cock of my own breeding. In 1955 I rung them all with split rings, violet for the original pair, blue and yellow for the other two pairs.

1. **The Yellow pair I put in the cage in the house and the first nest of three I had a large slate coloured hen and two Normals; after that I never seemed to have much luck with them.**

2. **The blue rung pair bred Normals with a few white flights and a white chin.**

3. **The original violet pair, bred me a good Pied, Normals and White-flighted Normals. Unfortunately I lost the hen and several young ones through knocking open the aviary door.**

In the 1956 breeding season on my friend's (Peter Pope) instructions I had purchased an unrelated Normal cock and mated it to my Silver hen and only reared seven young all season, not a Silver amongst them but a Pied hen which I will call (Pied 43) with some white chin and white flight feathers.

My cock of the Blue ringed pair died, so I mated the hen to the old Violet cock and I got some dark Pieds. I had a young cock off the old violet rung pair which I call V.10, I mated him to an unrelated Fawn hen and got four Silvers, three cocks and one hen. In the 1957 breeding season I mated my Silver cock to a Silver hen which resulted in five Silvers, one with a white chin. The old (Violet) cock I mated to a hen with a white chin (grand-daughter) and they produced a Silver hen, a White and some Pieds. Another round they gave a White, a Silver and a Silver Pied . . . It will be seen that all my Silvers came from the same source . . .

Although Silver Zebra Finches had been known in breeding aviaries for some years the first to be exhibited were at the National Exhibition in December 1948 when a pair were shown by S. N. Pinfield, but did not come in the cards in a mixed small seed-eater class of nineteen. It is interesting to note that in this same class a pair of

Fawns came Third and a pair described as split Silvers were of the Dominant kind and the split pair were Recessives and called split Silvers their colouring being a shade between Normal Grey and Dominant Silver. It will be realised that neither breeders or exhibitors were very conversant with these Dilute birds. It was quite a while before the Silver mutation was sorted out some saying they were Dominant and the other Recessive not being aware that two breeding kinds existed.

In the first Zebra Finch Society Year Book, Roy Wilson contributed an interesting article dated October 1952 where he said they were Dominant. In the same Issue A.J. Nash wrote saying that Silvers and Creams were being bred in South Africa but did not give their hereditary behaviour. In the 1955 Zebra Finch Society Year Book, Peter Pope contributed an excellent article on the breeding of Dominant Silvers and Creams which helped breeders to increase their stocks.

In 1954 The Zebra Finch Society issued their first set of *Colour Standards* which included Silvers and Creams. The colour given for Silvers " . . . From silvery grey to almost White . . . " and that the Creams ". . . from deep cream to pale creamy white . . . ". This was for both the Dominant and Recessive kind although no separation was given. These loose descriptions of the Dilute varieties held until 1965 when the committee of the Zebra Finch Society reviewed them in the light of the general advancement in colouring and gave a separate standard for both the breeding kinds of the two Dilutes. The colour for the Dominant Silver was given as being various shades of the Dilute Normals with silvery grey being the ideal and for the Dominant Cream as various shades of Dilute Fawn with an even cream being the ideal. The colour for the Recessive Silver was given as bluish grey and the Recessive Cream as medium cream. These descriptions gave breeders and exhibitors a good idea of just what was required for the colour of the two different mutations.

It was I think my wife and I who exhibited known Recessive Creams for the first time at the 1955 National Exhibition at Olympia, London. They were seen by many

Fanciers and there was a strong feeling that these birds were very much like very pale Fawns, whereas others thought them quite distinct from either Normal Fawns or Dominant Dilute Fawns. The history of this mutation is rather interesting as the first two birds (cocks) suddenly appeared amongst our Normal Grey breeding pairs. These two birds being cocks ruled out the possibility of Fawn being in their distant ancestry and we know their parents were Grey bred for some five or six generations. In the following season the two young cocks were mated to unrelated Grey hens and they produced a nice batch of Greys both cocks and hens. Further examples of these Recessive Dilutes were bred from mating the two cocks to young split hens. An unusual feature in the breeding of these birds was that no Fawns were bred from any of the pairings only Normal Greys and Recessive Creams. This seemed to indicate their colouring was due to a different character than the one that caused Fawns and Dominant Creams. Unfortunately, at that particular time we were short of aviary space owing to a series of colour breeding experiments we were carrying out and we had to dispose of this particular strain. We lost touch with the buyer so we do not know what happened to those particular birds or if the strain was perpetuated. Before this particular time I had received several reports from Australia, America, Europe as well as Great Britain of further Dilutes (both Creams and Silvers) appearing but no breeding particular were forthcoming. It will be realised that with any Recessive character it is always difficult to know if a sudden appearance is a new mutation or an old one that has been handed down unbeknown for generations and then suddenly appearing.

It would seem that Recessive Silvers are more plentiful in Europe and the birds bred there are generally of a more bluish tone than the ones we now occasionally see in this country. This fact makes it more understandable why they were called Blues to distinguish them from the fawnish tinted Dominant kind. The last specimen from Europe I examined was of an even slate blue shade with deep grey chest bar and tail markings with medium orange ear lobes and flankings a slightly lighter shade than found in Normals.

It was quite an attractive bird and I think that with selective breeding such a colour shade could be further improved.

The Zebra Finch Society's *Standard* for Colour for exhibition Dilutes is:

Dominant Silver: (Dilute Grey)
COCK: eyes dark, beak red, feet and legs pink. There are various shades of Dilute Normals (Greys) Silvery grey being the ideal. Chest bars vary from sooty to pale grey, cheek lobes vary from pale orange to pale cream, flankings from reddish to pinkish fawn with clear even white spots. Tear markings same shade as breast bar. The lighter the general colour the paler the chest, tail, lobes and flankings. Tail dark with white bars.
HENS: as other hens but of the same shade to match the cocks.

DOMINANT CREAM: (Dilute Fawn)
COCK: eyes dark, beak red, feet and legs pink. Again all shades from deep cream to pale cream. Markings in the cocks to be in general tone to match depth of diluteness. Tear markings same shade as breast bar. Tail deep cream with white bars.
HENS: as other hens but of the same shade to match the cocks.

RECESSIVE SILVER: (Dilute Grey)
COCKS: eyes dark, beak red, feet and legs pink. Head, neck and mantle medium bluish grey, wings grey. Throat and upper breast zebra striped, bluish grey with darker lines running from cheek to cheek continuing down to chest bar. Chest bar dark grey. Tear marks distinct and to match colour of chest bar. Cheek lobes medium orange. Underparts white, sometimes slightly shaded near vent and thighs. Flankings light reddish brown with clear white spots. Tail dark with white bars. Cock markings should be clear and distinct with only slight dilution.
HENS: as other hens but of the same shade to match the cocks.

RECESSIVE CREAM: (Dilute Fawn)
COCKS: eyes dark, beak red, feet and legs pink. Head, neck and mantle medium cream, wings cream. Throat and upper breast zebra striped, cream with darker lines running from cheek to cheek continuing down to breast bar. Cheek lobes medium orange. Underparts white, sometimes slightly shaded near thighs and vent. Flankings light reddish brown with clear white spots. Tail dark cream with white bars. Cock markings should be clear and distinct with only slight dilutition.
HENS: as other hens but of the same shade to match the cocks.

PIED GREY

Nearly all species of birds when fully domesticated produce a broken coloured mutation and they are generally called Pieds and sometimes Variegated as in Canaries. Before the actual Pied Zebra Finch mutation occurred and birds were bred that had one or two white feathers showing in their otherwise dark plumage. These birds with foul feathers did not reproduce their mis-marks as these were not due to a genetical mutation. However some time during 1927 a definite Pied form appeared in a mixed aviary of Zebra Finches in Denmark which is the same country that gave the Fancy the Recessive Pied Budgerigars. I consulted with my friend, the late C.af Eneljelm of Denmark, as to the precise origin of this Pied mutation but he was unable to obtain the necessary particulars. As far as records go this Danish Pied mutation is the only one to have appeared up to date and therefore all Pied Zebra Finches must have stemmed from this single source.

The first mention of Pieds as exhibition birds was at the October 1952 Swindon Show where a pair was exhibited by R.C.J. Sawyer who gained First place in the Any Other Colour Zebra Finch class. At the National Exhibition in November the same pair was placed Fourth in the A.O.C. class. In the following December, at a Zebra Finch Society Patronage Show held at Bristol, A.J. Wilkins was awarded Best Zebra Finch in Show with an exceedingly well matched pair of Pied Greys. This was the first time Pieds had gained such an award. Some of the early winning pairs of Pieds were very variable in their colouring and I remember a particular pair shown by the late Peter Pope of Ashford, Kent. These birds were of excellent type and substance but both had completely white heads which meant that they lacked tear marks and the cock was also minus cheek lobes. Such a pair of course would not meet present day standard requirements. It was not until 1954 that The Zebra Finch Society formulated a brief show standard for Pieds which was in force until 1961 when a more detailed one was compiled. However, after a few season exhibiting it was agreed that even the new Standard was not clear or precise enough. After considerable discussion in committee the present *Standard* was evolved. Much of the credit for this must go to A.A. Smith of Smethwick, Staffs, a Zebra Finch Society committee man and a very successful breeder of exhibition

Pieds.

Although Pieds had been in existence for some years they were first mentioned in an article in the Zebra Finch Society Year Book of 1954 where G.B. Shaw writes about Pied Greys and White-headed Silvers. This would indicate that Silver Pieds were actually in existence in the early nineteen fifties if not before. At this time it was not fully realised that the Pied character could be expressed in many forms of variegation. Once birds with the Pied character began to circulate amongst breeders it was not long before Pieds other than Grey Pieds began to appear in breeding aviaries and on the show benches. I now quote from an article by Peter Pope in the Zebra Finch Society Year Book of 1957 regarding the different colours in Pieds. He writes ". . . How many Pieds of different colours to the usual Grey and White does one see on the show bench? Until 1956 none, and then a pair of Silvers were benched in that year. I started to show a pair of Fawns at that time . . . ". Although these other coloured Pieds must have been bred for some years previous to 1955/6 but were evidently not exhibited. I feel certain that the advent of the Pied Fawns on the show benches helped very considerably to advance the breeding of all Pied kinds.

As the Pied character which is a Recessive became more widely distributed in breeders' aviaries so further Pied combinations came into existence. In the early 1960's I remember seeing examples of Pied Chestnut-flanked Whites both cocks and hens in aviaries containing mixed coloured Zebra Finches. Although a few examples were to be seen at bird shows I think most of these came from consignments which originated in Europe. A further Pied form that I believe also originated in Europe is the Pied Penguin in both the Grey and Fawn kinds. As far as I can recall neither Pied Chestnut-flanked Whites or Pied Penguins have found their way on to the show benches. Odd examples of Pied Lightbacks and Pied Black-breasted have also been bred but again have little to recommend them as exhibition types although they are interesting from the colour breeding angle. As I said earlier Pieds, have been used to improve strains of Whites which has resulted in Pied Whites being bred. Pied

Whites are of course identical in colour to Whites, masking any of the other colours and such birds can only be positively identified by test pairings.

When the *Standard* for Pieds, which is given below, is read it will be realised how difficult it can be to breed well matched exhibition pairs to the Standard specification. It is this challenge that has caused many breeders to specialise in the breeding of exhibition type Pieds in both their Grey and Fawn forms.

The Zebra Finch Society's *Standard* for Colour of Exhibition Pieds is:

PIED GREY:
COCKS, eyes dark, beak red, feet and legs pink. The grey colour to be broken with white approximately 50% of each colour. (White underparts not to be included in the 50% White). Cock markings to be retained in broken form on cheeks, flanks, tail and chest. Tear marks distinct but can be broken.
HEN, as other hens but of the same shade and broken areas to match cock.

PIED FAWNS:
as above with the grey colour being replaced by fawn.

PIED OTHER COLOURS:
There can be Pied forms of other colours and combinations following the same pattern as above.

PENGUIN

During the latter half of the nineteen forties and early nineteen fifties reports had been coming in from Europe that a new mutation had occurred that had different markings to those usually seen on Zebra Finches. The first details of these new birds was given to the Fancy through an article by the late C.af Eneljelm appeared in the 1955 Issue of the Zebra Finch Society's Year Book. The following quote from that article headed 'The White-bellied Zebra Finch'. From that article a clear picture can be seen of these new birds and their origin:

One of the latest mutations; as far as I can trace the least known, is

the White-bellied Zebra Finch. As far as I know they were first bred in Australia; the first ones reaching Europe came into the possession of the well known Belgian Fancier, Mon.L. Raymaekers of Brussels, from whose stock my birds originate.

The White-bellied resembles somewhat the better known Pied; but contrary to this the White-bellied is always rather regularly marked, variations being only slight.

The cock has silvery grey crown, neck, mantle and flights, the end of the big flight feathers usually being a bit darker. The underparts are pure white, rump and tail almost white. The Chestnut cheek lobes and flanks are maintained by the Zebra markings and black breast bar are absent. Colour of eyes, beak and feet are as in Normals.

The hen is like the cock, of course, without the cheek lobes and chestnut flanks. The cheek lobes are pure white. Usually the grey colour is somewhat darker than that of the cock.

The youngsters when leaving the nest are almost like Normals, the underparts however being more pure white as well as the rump. The beaks are black.

The first pair I got from Belgium in 1951 were rather on the small side but strong and healthy. They bred two youngsters only, a pair. Later from seven pairs I got many youngsters including some ten White-bellied . . .

It will be seen from the above quote that the mutation originated in Austalia the native home of Zebra Finches and that they were being bred in Europe not later than 1949/50. At first these birds were called White-bellied as indeed they are, but this description could also apply to other colour forms and in due course the very descriptive name of "Penguin" was assigned to them. This name is a good one as the birds have a completely white front from chin to vent similar to their namesakes. The first examples must have reached Great Britain somewhere about 1953/4 as specimens were exhibited at the 1955 National Exhibition of Cage birds held at the Olympia, London, by H.D. Porter of Smethwick. Penguins were recognised as a distinct variety by The Zebra Finch Society in 1958 but the *Colour Standard* was not published until it appeared in the 1961 Year Book.

The main source of information on Penguins came from C.F. Eneljelm through his articles in the Zebra Finch Society Year books. In the 1957 Issue he wrote:

There seems to be a lot of confusion as to the name. The name

'White bellied' was suggested by me some years ago, but I shall admit that the name is not a very good one, because most Zebra Finches are White-bellied even if the colour of the variety in question is more pure than in other varieties. Mr. Pope once suggested the name White-throated, which may be better as these birds of both sexes really have white throats – the cock being without Zebra markings – even if this characteristic might not be the most conspicuous one. On the Continent, especially in Holland they are now (1957) offered as 'Silverwings', and I believe that they have also been offered in England under this heading. I should say, however, that this name is not a very good one as it might cause confusion with the Silvers (Dilutes) . . . As to size and type the White-bellied seems by no means very easy to improve: they are still on the small side, with small narrow heads and large beaks . . .

In 1956 I mated some Normal/Fawn White-bellied cocks to White-bellied, Fawn/White-bellied and Cream/White-bellied hens. Also some Silver/Fawn White-bellied cocks to Fawn/White-bellied and Cream/White-bellied hens . . . The results of these matings were not very good; I got a lot of White-bellied but only two White-bellied Fawn hens and one White-bellied Silver cock. In due course these birds were mated to White-bellied and White-bellied/Fawns and from them I bred both cocks and hens White-bellied Fawns . . .

Both Normal Grey and Fawn Penguins do not show their full colour pattern until after their second and subsequent adult moults when the attractive lacing appears clearly on their wings. Although it is often a little dificult to match for exhibition any of the colours, it is even more so with the Penguins because of the wing lacings. A number of breeders keep and exhibit a few pairs of Penguins and in recent years the most consistent and enthusiastic breeders and exhibitors are N.H. and J. Cox, K. Ostermeyer, C. Scath, Mrs. J.Venner to mention a few.

It will be seen from Herr Eneljelm's notes that the original Penguins (White-bellied) were small, narrow headed birds and I think that this has had a definite bearing on their lack of popularity amongst breeders and exhibitors. When a colour mutation is linked to a special type it takes a long time and much perseverence on the part of breeders to alter the type and retain the colour. Judging from the present standard of Penguins seen at shows it is clear that their devotees have worked extremely hard in bringing the mutation to its present good standard. The system employed by some breeders was similar to that used by breeders of Danish Recessive Pied Budgerigars. Here they found that by using first cross Normal Greys (or Fawns) and mating them with their best Penguins there was a little improvement in the general quality of the Penguins ultimately produced. These Penguins were again paired with first cross birds resulting in a little further advancement. After several years of this breeding method the Penguins resulting were brought up to an excellent quality. To achieve such an improvement it is necessary for the breeder to produce a quantity of first cross birds each season by using good Normals and selected Penguins. By using certain strains of Normals which excelled in good deep chest bars some of the Penguins ultimately bred showed traces of chest barring which is so undesirable in this mutation. Bearing this point in mind it can be seen how important it was and is to pay special attention to the selection of each of the Penguins (and Normals) used in developing strains of these birds. It is interesting to note that in the Zebra Finch Society Year Book of 1969, W.B.C.Hocking wrote:— "I first bred Penguins in Normal form about fifteen years ago . . . The first birds I saw were either Continental bred or direct progeny and they had no trace of chest barring whatever and were pure white from chest to vent . . . "

The *Standard* of Perfection for Exhibition Penguins is:

NORMAL GREY PENGUINS:
COCK: eyes dark, beak red, feet and legs pink. Head, neck and mantle even silver grey, with flights, secondaries and coverts edged with a paler shade of grey giving a laced effect. (This lacing does not show to full advantage until the second full moult.) Underparts from beak to vent pure white without any trace of barring. Cheek lobes pale cream to match body colour of bird. Tail silvery grey barred with white. Side flankings reddish brown with clear white spots.
NORMAL GREY PENGUIN:
HEN: as other hens to match cocks but with cheek lobes white.
FAWN PENGUIN: as Normal Grey but with a soft fawn shade replacing the grey.

As yet Penguins have not won a **Best in Section Award** although all the colours previously mentioned have been successful in that respect. Given time the keen breeders of this mutation may achieve that honour with their steadily improving birds. Because of their unusual colour arrangements Penguins always attract the attention of visitors at bird shows and therefore are an added attraction to the Zebra Finch Section. During the period of the development of the Penguin mutation great events were happening in the Zebra Finch world. The most important one was that the committee of the Zebra Finch Society declared in the Fancy Press during 1958 that Zebra Finches were no longer to be considered as Foreign Birds but as a fully domesticated species like Canaries and Budgerigars. This declaration was welcomed by show promoting societies and they provided classes for Zebra Finches under separate headings. This led to extended classifications and special classes for current year birds. In the 1956 show season Junior exhibitors were given a special status and classes of their own at Patronage Shows. The declaration of domestication was quickly accepted world wide and Zebra Finches became known everywhere as fully domesticated birds. It will be seen from the following notice that appeared in The Zebra Finch Society's Year Book of 1958 that Zebra Finches had become firmly established in the Cage Bird Press. "In view of the new status of the Zebra Finch, Mr. J. Kuttner (Advertising Manager of *Cage Birds*) has agreed to the committee's request for a separate advertisement heading for Zebra Finches. Will members please mention this New Zebra Finch heading when putting advertisements in Cage Birds."

An early illustration of a pair of Zebra Finches by F. W. Frohawk

3

SOME INTERESTING VARIATIONS

"No other of the Australian Ornamental-Finches is so treasured and widely distributed as this, one of the smallest and most brightly coloured."

Dr. Karl Russ — *Pioneer in the breeding of Cage Birds and a leading authority in his day*

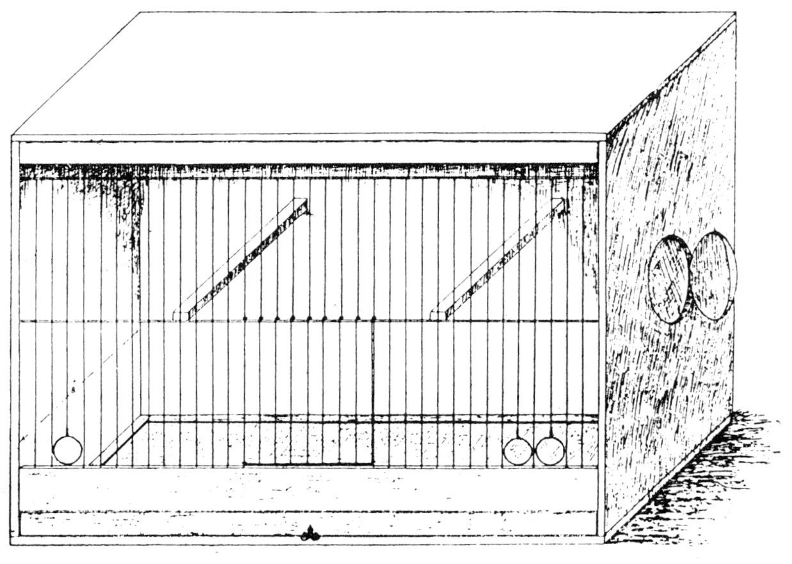

A Possible Stock Cage for Finches

CHAPTER 3

SOME INTERESTING VARIATIONS

YELLOW-BEAKS

It has been said by some field ornithologists that one of the numerous wild sub species of Zebra Finches has yellow beaks similar to the case of Grass Finches where one has Yellow and the other red coloured beaks. If this is correct then it is most likely that our present day Yellow-beaked birds are descendants from these birds. The Yellow-beak character is now known to be a Recessive one and as such can be carried hidden for many generations before revealing its presence. When Zebra Finches were first imported from Australia many years ago in. quite large numbers it is very possible that some Yellow-beaked birds were included in the batches of birds exported. Most of these would probably have been discarded as being poorly coloured specimens but nevertheless a few odd birds could have been passed through and when paired to Normal birds would only give normally coloured young. Although these birds would be normal in colour they would carry the Yellow-beak character in their genetical make-up. It is only when the Normal birds each carrying the Yellow-beaked character are paired together that a percentage of actual Yellow-beaked birds would be produced.

It is difficult to fix a date when the first Yellow-beaks were noticed amongst aviary breeding stock as being a different variety. However in the 1961 Zebra Finch Society's Year Book Stan Draycott of Intake, Doncaster, gave a report of how he had bred Normal and Pied birds with yellow beaks quite distinct from the normal red ones. During 1960 he paired a lightly Pied cock to a lightly Pied hen both birds being bred from two normally coloured parents. These birds produced two Normal Yellow-beaked hens, one lightly Pied Yellow-beaked cock and a number of Normals and Pieds. It can be assumed from this result that both the parent birds were carriers of the Yellow-beaked character as were some of the other red-beaked young from the same nests. Further Yellow-beaked birds were bred from the Draycott strain and became dispersed to Fanciers in different parts of the country. From the early nineteen-sixties onwards reports of the breeding of Yellow-beaked birds in various colours have come to hand indicating the spread of the Yellow-beaked character. In 1971 the committee of The Zebra Finch Society reviewed the question of the Yellow-beaked birds and decided to recognise them as a distinct form and formulated a standard of perfection.

Yellow-beaks in various colour forms have been bred in Europe in several different countries but I have not yet heard of any concerted efforts to produce strains of these birds. The first actual Coloured Plate of the Yellow-beaked variety that I can find is in an Issue of 'Die Gafiederte Welt' which was published in Germany in 1967 from a 1966 painting by H. Heinzel where a Normal Grey Yellow-beaked cock is depicted. This Plate shows the slight difference in the brightness of the grey colouring and the paler shade of the ear lobes which is quite in keeping with the several living specimens I have seen and handled over the past few years. In the *Colour Standards* published in 1968 by the Finch Society of Australia (Zebra section) Yellow-beaks are recognised as Yellow-billed. The *Standard* given is – **Cocks and hens: To conform to the appropriate colour variety except in each case the beak, feet and legs will be yellow.** This *Standard* is substantially the same as that in The Zebra Finch Society 1979 Handbook except that here the colour of the feet and legs is not given. The Zebra Finch Society's *Standard* reads **"Yellow-beaked Varieties. General colouring as Normal Grey and all other mutations except the beak which should be shades of yellow with the cock birds showing the richest colour. There can be a Yellow-beaked form of all existing mutations and their composite forms. Yellow-beaks must be exhibited in true**

pairs of the same mutation."

In the October 1982 Newsletter of the Zebra Finch Society, Jim Addison of Carlisle wrote an instructive article on the new Mutations. However amongst his observations on the mutations he said "Will there be a demand for the mutation? An example of this was the Yellow-beaked variety which proved to be a freak rather than a true mutation." Nevertheless in the February 1983 Newsletter Courtney Hocking of Redruth, Cornwall, contributed an article giving interesting details on his breeding over a number of years of the Yellow-beaked form in several colours. This information and numerous other facts already quoted from Fanciers both at home and abroad fully confirm that the Yellow-beaked birds are a true Recessive mutation. At the present time Yellow-beaked varieties in Great Britain are exhibited in the same classes as their normal red beaked counterparts. Yellow-beaks are last of the varieties on The Zebra Finch Society's list of standardised birds for show purposes. This of course does not mean that other mutations which now exist will not be included as and when sufficient numbers have been produced so that an exhibition standard can be evolved.

CRESTS

Once a species of bird has become full domesticated it is not only that changes in their colour eventually occur but also the arrangement of the actual feathers themselves. Such species as Canaries, Budgerigars, Bengalese, Pigeons, Poultry and Ducks, all have feather mutations mainly on the head in the form of a crest. The domesticated Zebra Finch has now followed the same pattern and a Crested mutation has occurred and has been fully established. I first received definite information that Crested Zebra Finches had actually been bred from that International Ornithologist and Artist Herman Heinzel. During one of our numerous conversations he told me that whilst visiting a city in Spain he saw in a bird shop two Grey Zebra Finches that had small round crests on their heads, these I gathered he obtained! At that time he was living in the Principality of Andorra where he maintained numerous aviaries of Zebra Finches and other small Finch-like birds for close scientific study.

In due course Herman Heinzel very kindly brought me a Crested Grey cock which fortunately proved to be a wonderful breeder and I think he is the ancestor of the majority of Crested Zebra Finches seen in Great Britain today. There may of course have been other Crested mutations but the one mentioned above is the only one of which I have definite knowledge.

The Crested character is a **Dominant** one and its production seems to be controlled to some extent by certain modifyers carried by the birds. Birds bred from one Crest parent mated to a pure non-Crest will produce a number of birds with crests of various degrees of perfection from the Full Circular to the Tufted. Being of a Dominant nature it is quite easy to breed Crested birds in all colours, varieties and their combinations, and up to date I have bred them in Grey, Fawn, White, Chestnut-flanked White, Light-backed and Black-breasted. It has been my experience that birds with the best looking circular crests are obtained from mating a Crest to a Crest-bred and it does not matter which of the pair carries the crest. **When two Crests are paired together more (about 75%) Crested birds may result, but some of these may have rather rough and ragged crests.** From the Crest to Crest-bred pairings about 50% Crests are produced with a high proportion of them having the desired Full Circular Crests, and of better quality. Crest-breds are birds that have been bred from one Crested parent and although they may look like Normals their breeding behaviour is different.

Writing of Zebra Finches in the *American Cage-bird Magazine* of October 1970, Dr. Val Clear of Anderson, Indiana, U.S.A., said ". . . Fanciers are always watching for new and exciting mutant specimens because this is part of the challenge of the Zebra Finch. A few years ago I thought I had bred a Frill but when I got a frizzled baby it died without producing any offspring . . . "

It was unfortunate that Dr. Clear's young bird did not

live long enough to reproduce but as such a bird has once occurred there is always the possibility of a similar feather mutation happening again or it may already have happened but has not been identified or reported. Although the Crested Zebra Finches have not yet been standardised the following *Standard* was suggested in the Zebra Finch Society Newsletter of October 1981. It will be seen from the following that the suggestion is an excellent one and may well be adopted.

CRESTED VARIETIES

Proposed *Standard* for Crests:

> **The same as set out for Normal Grey cock and hen and all the other different mutations (and their combinations) with the addition of a Full Circular head crest. The crest feathers should fall from a centre point on the top of the head forming a full circle finishing level with the top of the eyes.**

I have now discussed the Normal Grey and the seven Standard mutations also their Crested forms and will now turn to the non-standardised varieties that already exist.

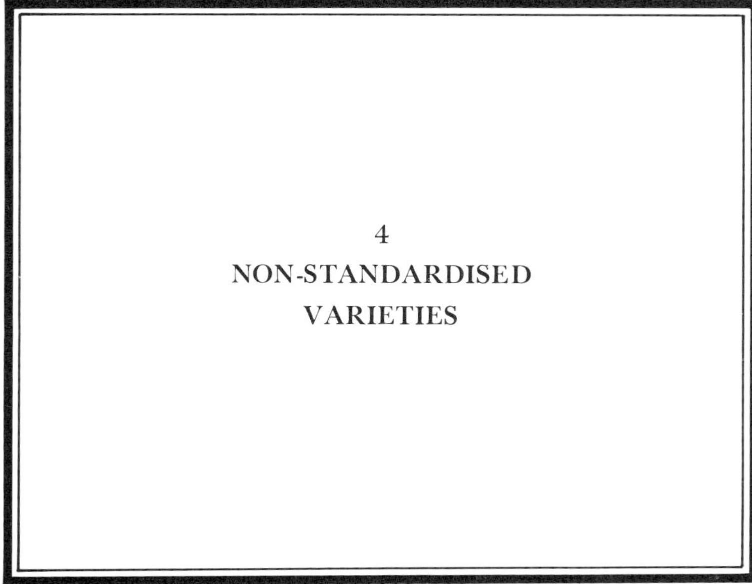

4
NON-STANDARDISED VARIETIES

A Melanistic Cock

Chapter 4

NON-STANDARDISED VARIETIES

LIGHT-BACK

This mutation was first noted and then investigated by Professor Dr. H. Steiner of Zurich, these birds were then known under the title of Hullruchen which of course translated into English means Light-back and it is by this name that they are now known. It must have been in the early nineteen-fifties when the Light-backs were first noticed amongst a batch of Normals breeding in a Swiss aviary. I did not learn of their existence until about September 1973 and it was not until 1975 that I saw the first live specimens that were brought to me once again by Herr Heinzel. In the following year I had further specimens arrive, this time from Franz Averstegge of Rosendahl, Germany, and also from breeders in Belgium and Holland. I now had three unrelated lines of these Light-backs with which to carry out experiments in discovering their full breeding behaviour. Other early breeders of Light-backs were Mr & Mrs Burns, D. Brown, Brian Binns, Mike Wrenn, Alan Pennington, Stan Draycott and Barry Deblin, whose breeding results helped in the investigation. Herr Averstegge contributed a most useful article on "Zebra Finch Mutations in Germany" in the 20th October 1979 Issue of 'Cage and Aviary Birds'. I now quote from this article ". . . The pale-backed (Light-back) mutation first appeared in 1954, in this variety the body colour becomes diluted especially on the back, but other marks are as normal. The mutation is sex-linked, but dominant to Chestnut-flanked White . . . "

It will be noted that Herr Averstegge stated that the light-backs were a sex-linked mutation and Dominant to the

Chestnut-flanked Whites. Breeding results in this country support this and most of the original Light-backs imported here were in fact carriers of the Chestnut-flanked character. At first I followed the breeding programme as suggested to me by Continental breeders and that was Light-back cock or hen to Chestnut-flanked White which gave 50% of each kind both cocks and hens. These pairings proved to be most prolific for myself and some other breeders who had related Light-backs, but in the beginning several found the number of cock Light-backs bred far exceeded the hens. Later matings seemed to settle down to the expected average of 50% of each sex.

Most of the early Light-backs bred in this country resulted from Light-back to Chestnut-flanked White crosses and it was Mike Wrenn who was one of the first British breeders to make the Normal Grey cross. I saw some of the Light-backs from these early Light-back to Normal Grey matings and they appeared to me to be substantially of the same colour as those from the Chestnut-flanked White matings. As Light-backs originally came from a mutation amongst Normal Grey birds such a result was to be expected. During the course of his experimental pairings with Normals and Chestnut-flanked Whites crossed with Light-backs Mike Wrenn noted that the Light-backs which carried both the Light-back and Chestnut-flanked White character were of a little lighter shade than those which had to Light-back characters in their genetical make-up. This point has now been observed by other breeders and myself when studying the colour results of our similar matings.

In the Zebra Finch Society Newsletter of February 1982 Brian Binns gave the breeding results from some of his Light-back pairings; amongst these matings were some Normal/Light-back to Chestnut-flanked White results and an unusual one from a Chestnut- flanked White cock to a Light-back hen which gave seven Light-back cocks three of which were Pied Light-backs. These were the first to be raised in Great Britain and to obtain such birds both of the parents must have been carriers of the Recessive Pied character. I think that the first time the Fancy had an opportunity to see living specimens of Light-backs was when I displayed several pairs at the Zebra Finch Society Club Show held in 1980 at Oxford. Since that time, numbers of pairs have been exhibited at Zebra Society Patronage Shows and in 1982 the Zebra Finch Society decided that for a period they could be shown in the Silver (Dilute) classes. Pairs have been shown at both large and small shows up and down the country where they attracted considerable attention both from exhibitors and the general public. Their pale grey upperparts and their normal dark markings of both cocks and hens make them a pleasing colour variety. When it was realised that the Light-back character was a sex-linked one their breeding created keen interest amongst those fanciers who liked to try new mutations and experiments were made to improve the variety show-wise with some success. The first coloured print of Light-backs I can find appeared in a Zebra Finch Supplement of the German magazine *Die Gafiederte Welt* of Summer 1967. A true pair of Light-backs were depicted along with a number of other mutations under the name of Hellruchen but it is not a good likeness as the upper body colour is far too dark. However it is the only colour drawing I have seen so far and it is interesting to note that it was printed in 1967 which is some years before we knew about them in Great Britain and a decade or so after they first appeared.

LIGHT-BACKS

A suggested *Standard* of colour for exhibition Light-Backs is:

COCK: Eyes dark, Beak coral red, Feet and legs reddish pink: Head light silvery grey with some darker ticking to centre of head. Neck, back and wings light silver grey sometimes with a light fawnish tinting. Throat to upper chest back black zebra lines on a white ground running from cheek to cheek. Chest bar black. Tear marks black. Cheek lobes medium reddish orange. Underparts below chest bar to vent white. Tail black with white barring, side flankings reddish brown ornamented with small round white spots.

HEN: as cock but minus chest markings, flankings and cheek lobes. Beak paler red. Some hens may have ghost cheek lobes.

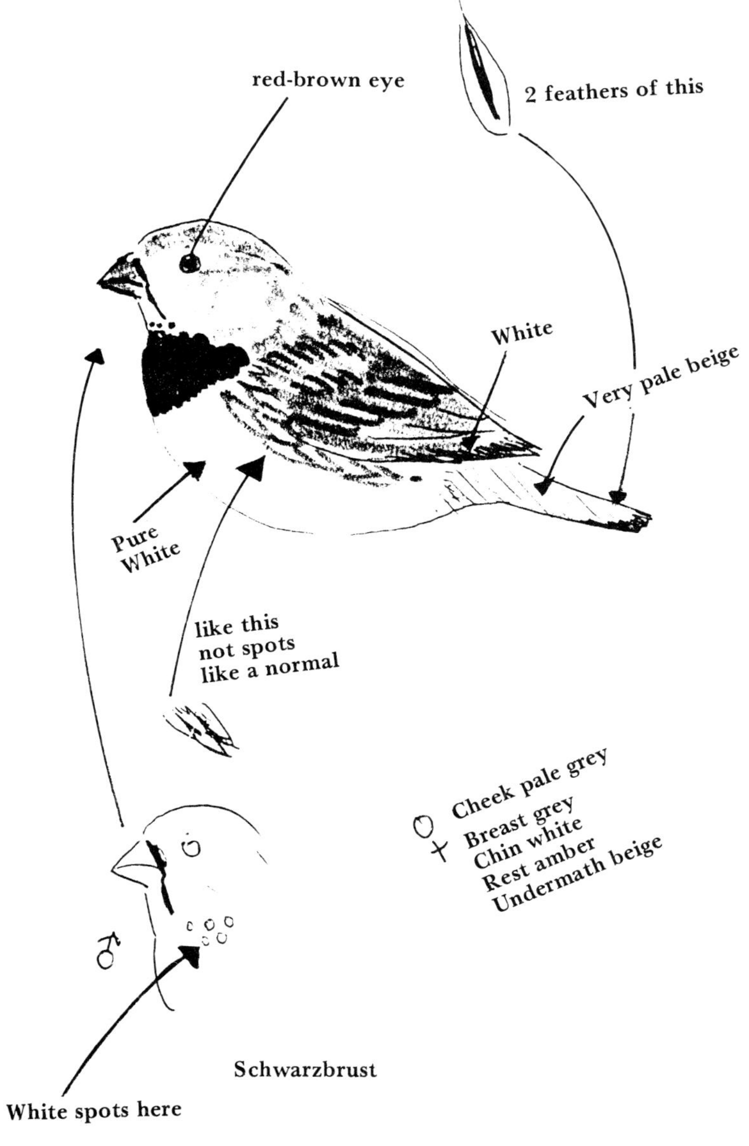

A sketch made by Herman Heinzel to show the colour formation

BLACK-BREASTED

It appears that two different mutations of Black-breasted Zebra Finches are in existence both with similar but not quite identical pattern markings, one turning up in Australia and the other in Germany. I have not been able to ascertain just when or where the German Black-breasted or Schwarzbrust as they call the mutation occurred, but it must have been in the mid-nineteen-sixties judging from my correspondence of that period. I was told that there was a high mortality amongst the cock birds in the early days of their development. When I started to breed them I did not fortunately have this experience, the inherent weakness must have been bred out by the process of crossing the German stock with the British bred birds. Through my letters from Europe I knew something about the Black-breasted kind but it was not until I discussed them with Herman Heinzel at a National Exhibition at Alexander Palace, London, in 1975, that I got a full description of their colour arrangements. On page 48 is pictured the sketch he drew from momory in his notebook and it will be seen that this is an excellent description of the mutation when compared with live specimens. Although at the time I was not aware that the prolific Crested Grey cock Herman Heinzel brought me in 1974 was a carrier of the Recessive Black-breasted character. In the course of breeding birds produced by this cock (Crested Grey) half brothers and half sisters were paired together and from some of the numerous offspring from the related pairs a few Black-breasted cocks and hens appeared some of which were Crested. These were the first that I had bred and were the foundation of the Black-breasted strain I have today. During the course of the breeding a few other Black-breasted specimens were imported into Great Britain so I am not quite sure who was the first in this country to produce the mutation.

I also imported Black-breasted cocks and hens from Germany including a Black-breasted Fawn cock, but unfortunately he was not a reliable breeder and further Black-breasted Fawns were not bred from him. Several of my Black-breasted birds turned out to be split for White and gave me a number of Black-breasted Whites. This White form is quite interesting as they show varying amounts of black edging to their flight and secondary feathers. It may well be that by selective breeding from these Black-breasted Whites that a strain of Whites having black lacing can ultimately be produced. In the Zebra Finch Society's *Silver Jubilee Handbook* the progress of the various Zebra Finch mutations are given and the Black-breasted are included amongst the newer varieties. At that particular time their genetical make-up was not fully understood but it has now been completely cleared up showing that the Black-breasted character is a Recessive. I think that the initial misunderstanding of their inheritance was due to the fact that some Normal birds bred from one Black-breasted parent showed certain Black-breasted characteristics such as lack of tear stripes, irregular shaped flank spots and blotchy tail barrings.

The Black-breasted colour pattern can be had in other varieities and I have some examples in Fawn, Chestnut-flanked White, Pied and Crested. A few pairs of Black-breasted birds have been seen at some of the Zebra Finch Society Patronage Shows with the first being at the South Western Zebra Finch Club Area and the Zebra Finch Society Club shows during 1980 and at the National Exhibition also in that year. The progress and popularity of Black-breasted is likely to be slow as breeding stock is very limited and being a Recessive it makes their production more difficult than the Sex-linked and Dominant varieties.

Another Continental Coloured Plate of the 1960s I possess, depicts a mutation called Shwarzling which has some characteristics of the Black-breasted, but is not known in this country as far as I can trace. The cocks have a complete black breast from chin to breast bar, black flecking on mantle and some black flight and secondary feathers, black at top of legs and the flanking has only a very few white dots. The tail is barred black and white like in a Normal Grey and the cheek lobes are of the usual size and colour. The hens have black flecking on the mantle and black at top of legs. I have not yet been able to find any particulars of their breeding behaviour or even if the birds still exist and I am wondering if such birds may have been the forerunners of what we now know as Black-breasted. In Australia a further variant(?) of the Black-breasted has appeared which they have named Black-fronted. The colour *Standard* for these birds given in The Zebra Finch Society of Australia show *Standards* as I quote: **"As for Normal Grey, except Face from Tear Drop to Beak 50% Black, Breast Bar Black, extending up into Upper Breast. Tail black, devoid of any White bars."**

From this description it can be seen that these birds have more black colouring in their plumage than our Continental Black-breasted and would appear to be a different mutation. Amongst the same set of Show *Standards* a further mutation is given and called Black-faced and is described as follows: **"As for Normal Grey, Black Tear Drops extending to Beak, side flanking Dark Reddish Brown with clearly defined White Spots. Breast Bar Black extending down as far as possible towards the Vent."** The black colouring of this variety goes the opposite way to that of the Black-fronted and the whole facial area is black.

I first had information about the existence of the Black-faced variety in 1967 from I.C. Sharam of Griffith, New South Wales. In 1969 he kindly sent me fuller details of this mutation and I quote from his letter:

> **. . . I now have some further results from pairing the Black-faced birds. From the mating of four Black-faced Zebra cocks to four wild caught Grey hens I had three Normal Grey cocks, two Normal Grey hens, two Black-faced cocks and one Black-faced hen.**
>
> **The mating of a Black-faced hen to a Normal Grey wild caught cock produced two Normal Grey hens, one Normal Grey cock and one Black-faced cock.**
>
> **We have now mated Black-faced cock to Black-faced hen which gave three young, two Black-faced hens and one Normal Grey cock. The amount of black varies with each individual cock bird as in Pieds in Budgerigars. There are now some thirty odd Black-faced cocks and eight hens in this area.**

Australian breeders are very lucky in the fact they can get pure bred wild Zebra Finches for carrying out experimental test pairings. From these breedings results just mentioned it can be assumed that the Australian Black-faced mutation is a Dominant one as both cocks and hens paired to Normal Greys have given a percentage of Black-faced young. When two Black-faced birds were paired together both Black-faced and Normal young resulted showing that the new character was visible in a single quantity as would be expected with a Dominant. Over the past fifteen or so years I have had reports from several breeders in this country of Black-faced young appearing amongst their young stock. Some of these specimens assumed normal plumage after a second moult and the few that retained the black face did not live long enough to produce any Issue. I also bred a few Black-faced young including a Brown-faced Fawn but all these birds died within a few months. There are now three established breeding strains of Zebra Finches with an excess of black in their feathers – the European and Australian Black-breasted and Black-fronted and the Australian Black-faced.

In 1977 I had a report of a bird with blackish ear lobes instead of the normal orange ones being bred from a pair of Ordinary Normal Greys. This bird also had very dark brown flankings with its grey colouring being darker and duller than its grey parents. A year or so later I saw and eventually obtained a similarly coloured cock of unknown but obvsiously very mixed parentage as he proved to be split for Fawn, Pied and White.

In his first season paired to a Normal Grey hen they only raised to maturity one chick which turned out to be a White cock. The following year I mated him to another Normal Grey hen and this time the pairing was blessed by seven young in the following colours – three Pied Fawn hens, two Grey Cocks, a Grey hen and a Pied Grey hen. The next season I put the old Grey-cheeked cock and his two Grey sons to the three Pied Fawn hens. From these three pairs I got fourteen young consisting of three Grey-cheeked cocks, one Fawn cock, one Grey cock, three Pied Fawn hens and six Grey hens. Four of the Grey hens showed slight variations from ordinary Greys and I took them to be Grey-cheeked hens not having seen one previously. Before the following season two of the Grey-cheeked cocks and three of the hens died leaving the old cock and one young one. As the breeding season begins I have three cocks and four hens ready to start and I have great hopes of being able to increase the numbers of this mutation.

If I am fortunate enough to establish this mutation I suggest the following Colour *Standard* could be adopted – **Grey-cheeked cocks: eyes dark, beak red, feet and legs dull reddish brown. Head grey ticked with darker grey on top. Neck, mantle, back and wings even deep grey. Under-parts below bar white. Side flankings dark brown ornamented with small white spots. Cheek lobes grey, tear marks black. Chest bar black. Throat and upper breast above bar black zebra stripes on a white ground, running from cheek to cheek. Tail black with white bars. Hens similar to cock but without markings except tear marks and tail barring.**

BLACKS

Undoubtedly the all-Black Zebra Finch has been much tried for during the past few decades and especially so when actual Black specimens have appeared in nests. Most of these Black Zebras have died before they could reproduce or in the process of moulting they have gradually reverted to Normal much to their owners' disappointment. The first mention I heard of the possibility of Blacks came from the late Peter Pope of Ashford, Kent, who was breeding from a strain of Greys that produced some birds showing extra black colouring in their plumage. In 1956 my friend, the late T. C. Charlton of Holland-on-Sea, Essex, actually bred one all black cock and three other birds pied with black all in one nest. The parents of these were a Normal Grey cock and a Fawn hen. It is interesting to note that both the Pope and Charlton birds were bred before the established Black-breasted and Black-faced forms were produced. The following extract from the Zebra Finch Society's *Year Book* of 1957 gives a picture of what was happening regarding the breeding of Blacks at that period:

. . . During the past few seasons a considerable amount of

interest has been taken in the production of a Black Zebra. Mr. Peter Pope has been working hard with a particular strain and has had a certain amount of success. Our Finnish Member, Mr. C.af Eneljelm, reports in this Bulletin of a Black Zebra being shown in Denmark. I am now pleased to say that our member, Mr. T.C. Charlton, has had a melanistic bird occur with one of his pairs last season. This pair produced three birds – two showing a large amount of black and one a complete Black and one further bird with a small amount of black. The Black bird, a cock, is all black except for dark chestnut cheek lobes and flankings, these flankings are devoid of the usual white spots. Mr. Charlton has mated this Black and is now awaiting the result . . .

I saw the Charlton Black a number of times and the bird certainly showed nothing but black and very dark chestnut brown in its plumage, not a trace of white anywhere. He was paired to a good deep coloured Grey hen and although the pair had two clutches of eggs they all proved infertile. Unfortunately, this Black and his black-marked nest mates all died without any issue.

I gathered that the Danish Black also died without producing any offspring. Peter Pope's melanistic strain did not give any actual black-feathered birds and eventually died out. During the nineteen-sixties and seventies further Black and partially black feathered birds were bred almost annually but no strains were founded from them. Any birds with black feathers that actually reproduced after they had moulted their abnormal feathers assumed dark grey plumage and all the young were ordinary Normals. It would appear that the 'black' colour was not of an hereditary nature and must have been due to a quirk in their metabolism. An unusual bird was reported from the Birmingham area in the early nineteen-eighties which was completely black on face, throat, chest and underparts down to the vent where a few white feathers were showing. The head, neck, mantle and wings being grey except for the cheek lobes which were deep orange but the side flankings were completely absent. It was not reported whether this bird actually raised any young. A further bird with solid black underparts was reported from the Birmingham area in the early Eighties but no particulars have yet been obtained and it is not known if the bird survived and was used for breeding. Last year yet another bird was bred which had a black face and a lot of black on its throat and chest and no white spots on its dark chestnut flanking. I gathered that this bird was bred in a mixed coloured collection of Zebras and that there were others of somewhat similar colour but with a little less black in the same nest. If further information comes to hand before this book is finished I will give details.

Special Note: Dr. Klaus Immelmann (*Australian Finches* Angus and Robertson, Sydney) quotes Professor Steiner of Zurich in asserting that the "melanistic" condition is due to a lack of Vitamin D in the diet.

ALBINOS

From the birds with black plumage we now go to the other extreme that is birds where all the dark colour is missing. Although a number of pure white birds with red eyes (Albinos) have appeared in this country strains have not so far been established. However the mutation that occurred in their native home of Australia have been fully established and examples appear regularly at their shows. I have not been able to establish just when or where this mutation turned up although I think it must have been during the 1960s. Another was reported coming from India again appearing in an aviary of mixed colours that had been breeding together without fresh blood for some years. Undoubtedly, under such conditions cross pairing and close inbreeding took place and this appeared to result in red-eyed clear Whites appearing. These Albinos were of very poor physique and many died in the nests and it is not known if it was possible to establish a strain. Odd Albinos have been bred in several European aviaries but again unfortunately strains from them do not seem to have been evolved. Albinos are invariably the result of a sex-linked mutation and the first Mutants are generally hen birds. An Albino hen paired to a Normal cock would produce all normally coloured young and it would be the cocks only that would carry the new character. These Normal Albino carriers paired to Normal hens would give the expectation of 25% Normal hens, 25% Normal/Albino cocks, 25% Normal cocks and 25% Normal hens. To breed Albino cocks a Normal carrier must be paired to an Albino hen which gives the expectation of 25% Albino cocks, 25%

Normal/Albino cocks, 25% Albino hens and 25% Normal hens. It always appears more difficult to establish a strain when the original mutant bird is a hen as everything depends on this single bird. There is a far better chance of success with a Mutant cock and he can be mated to a number of different hens thereby extending the chance of raising young birds. I should add here that an Albino cock or hen can mask any one of the other established colours.

ALBINO

Cock: beak coral red, eyes clear red, feet and legs pink, all feathers pure white without any markings of any kind.

Hen: same as cock but with a little paler beak.

Bicheno + Zebra Finch Hybrid. Some interesting Hybrids have been developed (see page 63/65)

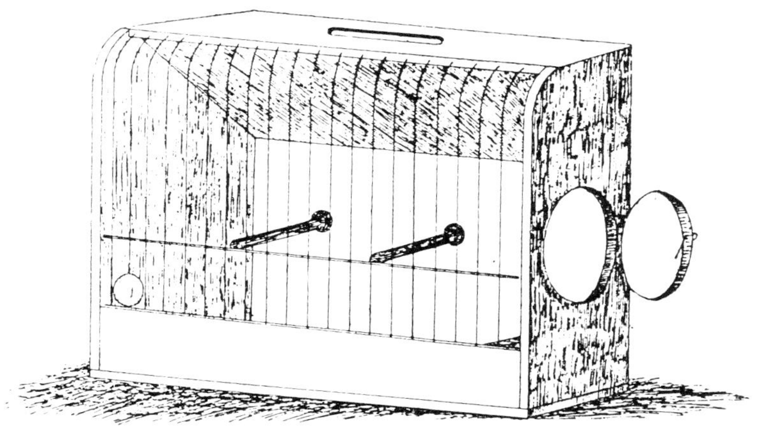

SMALL FINCH SHOW CAGE 12" x 10" x 5".

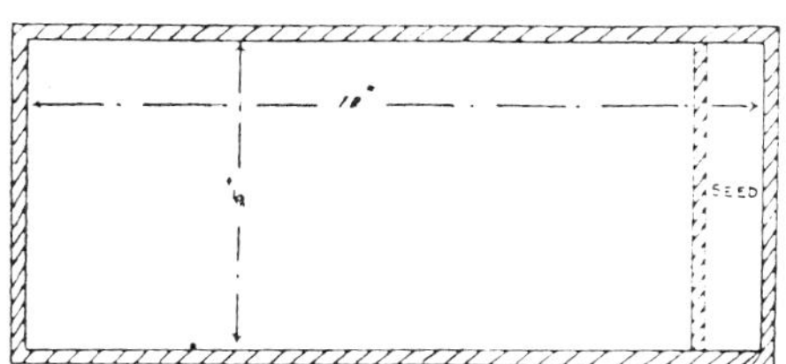

PLAN OF SMALL FINCH SHOW CAGE.

GRIZZLES

This mutation is yet another one which originated in Australia and is not known as far as I can discover in Europe although a breeding pair were in the possession of J.A.W. Prior, Secretary of The Zebra Finch Society some years ago. Unfortunately, these birds died without producing any further Grizzle issue. It is difficult to ascertain just when the first grizzle birds appeared although it is thought to have been in the early nineteen-seventies. They seem reasonably well known in Australia and the Zebra Finch Society of Australia recognises them and have formulated an exhibition standard which I shall give later. The grizzled feather effect is to be found with other domesticated species of birds and animals such as Pigeons, Poultry, Cats, Dogs and Rabbits to mention some of the most frequently met with kinds. In the case of Grey Zebra Finches the whole of the grey colouring is heavily flecked with white giving an unusual pepper and salt look to the feathers. From the scant details I have been able to obtain of this mutation I gather they belong to the Recessive group and they reproduce in that way and can be had in all the dark colours. As far as I am aware no other similar mutation has been reported as occurring elsewhere although it is possible it could appear at any time.

Last year however I had a number of reports of several Normal Greys and one Fawn appearing in widely separated parts of the country that carried considerable white flecking on their heads and necks and strange to say they were all hens. I have two such hens which are paired to Normal Grey cocks and any young cocks bred I intend to pair back to the adult hens next season in a hope that the area of flecking in some of their young may be extended and in the end cover the whole body. The cheek lobe area of my two hens show white flecking and it will be seen from the Australian Standard for Grizzles that cock birds have only a small amount of orange on their cheek lobes which is rather an unusual feature with cocks.

GRIZZLES

The Australian *Standard* of colour for exhibition Grizzles is:

COCK: to confirm to the appropriate colour variety, but appearing to show a white fleck on all feathers, thus creating a greyish effect; i.e., Grizzled. The cheek lobes to be body colour with approximately 12% orange shown near the tear drop.

HEN: as for appropriate Colour Variety, showing a whiste fleck on all feathers. Hens will show a cheek lobe of lighter body colour.

Saddle-backed Cock

SADDLE-BACKED

For quite a number of years now reports of the breeding of White birds with grey or fawn coloured saddles or mantles have been circulating in the Fancy. On close examination of the birds themselves and their pedigrees it is evident that they are heavily flecked Whites and therefore do not belong to a separate mutation. I think it may be possible by careful selective breeding to constantly produce these White birds with distinctive saddle markings. Amongst some of the birds I have seen have been saddle marked White birds that have been bred from Black-breasted stock where the flecking is black as compared with the grey or fawn of the ordinary birds. Some of these birds also have black edging to their flight feathers and have been proved by breeding tests to be the White version of the Black-breasted. Again it should be possible by selective breeding to further increase the amount of black and get a black and white variation. From what I have gathered from my European friends Saddle-backs, possibly including the Black-breasted kind, have been bred in several countries. Although I have not actually seen specimens of these birds their written descriptions seem to tally with the birds we have in this country. It will be interesting to see during the next few breeding seasons if there will be any further developments with this type of marked White bird.

A year or two ago I had details from a West-country breeder here in England that a Saddle-backed cock had been bred which showed most of the usual cock markings. The breeder kindly sent me an excellent coloured drawing of the bird showing clearly its pattern markings. All throat, chest and tear marks were absent but the orange cheek lobes, flanks and black and white tail stripes were present on a white ground with a deep cream mantle from nape of neck to beginning of rump. Unfortunately I have not heard any further particulars about this bird or whether he produced any offspring. A variety patterned like this would be most attractive and I am sure would appeal to many Zebra Finch Colour breeders. On comparing this bird with the mutation I shall be discussing next, shows there are many similarities in their colouring. In fact they are almost identical except for the saddle marking. If in the course of time a Saddle-backed variety is evolved it could be just like the bird mentioned above or a White bird with a dark saddle and full normal markings. As the variety does not yet exist a suggested Standard is not really applicable.

FLORIDA FANCY WHITES

This is an American mutation that has recently appeared, but unfortunately live specimens have not yet arrived in Great Britain or as far as I know in Europe. The first information I had that this mutation existed came from W. L. Cotta of Glendale, California, U.S.A. in 1977. Later I had supporting details from both T. Dunham of St. Petersburg, Florida, and Dr. Val Clear of Anderson, Indiana. The original mutant stock appeared in 1976 in the aviaries of Mrs. H. Kipp of Florida, U.S.A. which contained a mixed collection of Whites, Chestnut-flanked Whites, Fawns and Dilutes. In a further letter W. L. Cotta writes and I quote:

> **. . . M. M. McHarg writes that they have uncovered an Australian Fancy White that the owner had bred since 1971 but did not realise they were a new variety, until he read a report and description of the Florida Fancy Whites. . . . Apparently if my information is correct the American Florida Fancy Whites and Creams are not the first mutation of this kind . . .**

From the above notes it will be seen that although Florida Fancy Whites were first recognised as a mutation in America during 1976 a similar one had been bred in Australia some five years earlier. As these birds look something like Chestnut-flanked Whites it is quite understandable how they could be over-looked in a large mixed coloured breeding colony of Zebra Finches. This being so, it is quite possible that the mutation had appeared previously and had been passed unnoticed. My correspondence indicates that the Florida Fancy Whites like the Chestnut-flanked Whites exist in two colour forms – the near white and the pale cream kinds. Up to date I have not been able to establish for certain if they are sex-linked or Recessive or how they behave when paired with other established mutations. As Chestnut-flanked Whites were amongst the original stock from which the American Florida Fancy Whites were derived and their similarity in a number of colour respects I first thought there was every possibility they could be a sex-linked variety. Later details from W. L. Cotta show that this may not be the case as both cocks and hens have been bred from birds not showing the Florida Fancy White colouring. Unless we have a like mutation in Great Britain or Europe it is likely to be a very long time before we see specimens on our show benches.

FLORIDA FANCY WHITES AND CREAMS

A suggested Standard for this mutation is:

COCK: Eyes dark; Beak coral red; Feet and legs, pink; Head, neck, wings, throat, chest and underparts near white or pale cream. Cheek lobes bright orange. Flankings rich chestnut ornamented with round white spots. Tail barring very pale fawn.

HEN: As cock less all markings except for ghost tail barrings.

ORANGE-BREASTED

In the early part of this decade I had information from my friend Nils Svensson of Sweden of a new Zebra Finch mutation that was being developed in Belgium and because of its colour has been named Orange-breasted. This mutation is somewhat similar in overall colour to the Normal Grey with the exception of the black chest bar which has been replaced by an orange coloured bar of the same shade as the cheek lobes. This change of colour of the chest bar gives cock birds a striking appearance wuite distinct from all other mutations. As hen birds do not have a chest bar they will look more or less like ordinary Grey hens. It would appear that the first mutant, a cock bird, was bred in 1978 and that the colour is thought to be Recessive in its breeding behaviour. At first the production of the Orange-breasted has been rather slow but undoubtedly as the strain is developed, specimens will become more readily available. They have not yet been seen in Great Britain.

GREY CHEEK LOBED

The number of Zebra Finch mutations seems to be for ever on the increase with the latest news coming from France where birds showing a change of colour of their cheek lobes which are overlaid with a leaden (grey) suffusion have been bred. I have not yet seen living specimens but the coloured photographs and slides I have examined show a bird of the Chestnut-flanked White type with a heavy dull grey wash over the cheek lobes and very faint fawn markings on chest, flanks and tail, on a near white ground. Another photograph shows a bird with a little heavier markings and the grey suffusion on the lobes more restricted in its area. From these colourings it may mean this grey or lead as the breeders call it, colour can be had in two strengths. Double strength gives the full overall (grey) colour and the single produces a more restricted and lesser suffusion. This is of course assuming, as I do not know just how this mutation rerproduces or whether it can be had in other mutations than the original Chestnut-flanked White form. Hen birds of course lacking the distinctive cock markings would appear as Chestnut-flanked White hens with faint tail barrings.

In some of the previous pages on the Black-breasted Mutation I described a further Grey form I have that has a heavy grey suffusion on the cheek lobes. Here again I have not yet ascertained just how this Grey-cheeked mutation reproduces or if it can be had in other colours. I am now wondering if the French Grey-cheeked Chestnut-flanked Whites and my Grey-cheeked Greys owe their colouring to the same mutation. If in due course of time they are proved to be the same then it should be possible to breed a whole range of colours with grey cheek lobes and other associated slight colour changes. One of my Grey-cheek lobed cocks paired to what I think is a Grey-cheek lobed hen has so far given four chicks three of which are Grey hens and one a White cock which has a pronounced saddle. Being a White bird, cheek lobes are not visible so although a cock I still cannot tell visually if he has the Grey-lobe character. A son of the Grey-lobe cock of the pair just mentioned mated with a Grey-lobe (?) hen also had four young, two Grey hens, one Grey cock and one White hen, again with a pronounced grey saddle. The young cock would appear to be going to develop into a Grey- lobe although this strain produces many more hens than cock birds. The answer to this will no doubt be revealed in due course of time when more birds have been bred.

ZEBRA FINCH HYBRIDS

The number of new Zebra Finch mutations has been

quite considerable over the past years and they are now likely to become few and far between as there are limits to what can be done with the existing colours. There is of course always the possibility of a rare mutation giving a new colour phase which could be transmitted to all the existing varieties. However, it may be possible to create new colours and new pattern markings by means of fertile hybrids if such birds can be produced. In the Canary world the attractive Red Factor range of shades came about through a fertile Hybrid. On looking at the entries in Foreign Hybrid classes at National Exhibitions for the last fifty years or so I find that a number of Foreign Zebra Finch Hybrids have been shown. Their numbers may have been greater than the catalogue records show as numerous birds have been entered without giving their Hybrid parentage. In addition to the specialised breeders that are exhibitors there must be a considerable number of other breeders in whose aviaries Hybrids have appeared in their mixed collections of small Foreign birds where Zebra Finches have been present.

The examples shown at National Exhibitions include specimens of the following crosses with the male parents being indicated first. African Silverbill x Grey Zebra Finch, Chocolate and White Bengalese x Grey Zebra Finch, Bicheno's Finch x Grey Zebra Finch, Diamond Sparrow x Grey Zebra Finch and Grey Zebra Finch x Indian Silverbill. Some of these specimens of Hybrids have been produced and exhibited on a number of separate occasions. I have discussed the production of some of these Zebra Finch Hybrids with their breeders and they told me that the Hybrids came from aviaries of mixed small Foreign birds that contained the odd Zebra Finch. Under such circumstances single birds of opposite sex will often pair off, nest and sometimes raise young. To successfully breed Zebra Finch Hybrids from selected crosses the birds will have to be housed well away from the sight and hearing of other birds of the same species. If success colour wise is to be achieved the most important step will be to find suitable small foreign birds that have different colours or patterns to Zebra Finches and at the same time are birds that could possibly produce fertile Hybrids. In the case of the Red

Factor Canaries the initial crosses with Hooded Siskins produced Hybrids where the cocks only were fertile and the same procedure could be the same with any Zebra Finch Hybrids that are raised.

Possible species to use for crossing which have red in varying amounts in their plumage and belong to the same wide group as Zebra Finches are the Fire-tail Finch, Red-eared Fire-tail Finch, Painted Finch, Crimson Finch, Sundevail's Waxbill, St. Helena Waxbill, Black-bellied Fire Finch, Vinaceous Fire Finch and Blue-billed Fire Finch to name the most likely to be successful. Most of these species are quite rare in this country and Fanciers having pairs are most likely to use them for straight breeding rather than the somewhat dicey business of hybridising. However, there may be a number of odd birds of either sex of these varieties that their breeders may be willing to use for hybridising rather than just leave them to be decorative. Anyone having a spare bird would be well advised to contact a Zebra Finch breeder who is interested to embark on an unusual colour breeding programme.

There are not many small Foreign birds that have a yellow ground colour which gives green colouring although one does spring to mind at once – the Green Avadavat. Two futher possibilities are the Green-backed Twinspot and the Green-breasted Waxbill, both of which might make suitable mates for Zebra Finches. It will be interesting to see in due course of time if any of the above mentioned birds or other species will be tried out in an endeavour to produce the first cross Zebra Finch Hybrids.

In 1983 I had a letter giving details of two cock Zebra Finches of unusual colouring; these birds were minus tear marks and breast bars with light grey heads, wings and backs, mainly black tails and a couple of white splashes on wings and cheek lobes of light red. It is the appearance of the red colour in the cheek lobes that is exciting and could give rise to something quite new. This season's breeding results from these birds have not yet come to hand so I do not know if this red colouring has been perpetuated. A few weeks ago I had a *Newsletter No.56* issued by The Zebra Finch Society of Australia and amongst the various items is an article and sketch of what is described as a 'red Zebra Finch'. This name may be rather ambitious but according to the descriptions there is certainly some red colouring in their plumage. Having just written about the possibility of getting fertile Hybrids to try and introduce a red colour into Zebra Finches I now have information from Australia that birds with red in their plumage have already been produced. It would seem that these Zebra Finches originated in a Black-faced strain which are not yet known in this country. The following extract from The Zebra Finch Society of Australia's *Newsletter No.56* of 1984 gives the extent of the red colouring so far achieved by Greg Carey the original breeder:

> **There were two hens and three cocks. Two of the cocks were plain normals, the third cock was also normal, but instead of a black bar the bar was the same colour as his flanks.**
>
> **All hens that I have bred this year, carry the pink ear patch, some more pronounced than others.**
>
> **The ear patch of the cocks are slightly brighter than normal, and also bigger. (The larger cheek-lobes are also a feature with the German Black-breasted form.) There is a reddish tinge also, on the head feathers, on some of the cocks. Some show red above the bar and down the front. The back and wings are the same grey colour as in the black-faced Zebras.**

It will be seen from the above quote that these birds do appear to have a definite red or pink colouring which can be inherited and seems to be increasing in its area. If the increase is maintained and extended it should be possible to produce Zebra Finches with large areas of red colouring. What intrigues me is that this red colour has appeared in birds that have extra black in their plumage. Bearing this in mind one would be lead to think that black would be the colour to increase its area but not so in this instance. This illustrates how complex the feather colours in birds can be and one does not know what will be appearing next. The appearance of the red colouring brings up to date the long list of colour variations so far reported in the plumage of Zebra Finches. In the future there may be other mutations and colours and especially so if any fertile Hybrids can be produced and will breed. With all this large range of colours it can well be understood why Zebra Finches have become such popular cage, aviary and exhibition birds world wide.

All the different mutations follow their own particular method of inheritance which can be Dominant, Recessive or Sex-linked and the sets of rules given below will show how each kind operates.

DOMINANT INHERITANCE

Single character Dominant colour paired to a Normal gives 50% single character Dominant and 50% Normal.

Double character Dominant colour paired to Normal gives 100% single character Dominant.

Single character Dominant colour paired to a single character Dominant colour gives 25% double character Dominant colour, 50% single Dominant colour and 25% Normal.

Single character Dominant colour paired to double character Dominant colour gives 50% Double character Dominant colour.

There is no visual difference between the single and the Double character birds; this can only be ascertained by test breeding. It should be noted that no Normal colour can be split for any Dominant characters.

RECESSIVE INHERITANCE

Double character Recessive colour paired to Normal/Recessive character gives 50% Double character Recessive colour cocks and hens and 50% Normal/Recessive colour character cocks and hens.

Double character Recessive colour paired to Normal gives 100% Normal/Recessive colour character.

Normal/Recessive colour character paired to Normal/Recessive colour character gives 25% Double character Recessive colour 50% Normal/Recessive colour character and 25% Normal.

Normal/Recessive colour character paired to Normal gives 50% Normal/Recessive colour character and 50% Normal.

Whereas with the Dominant colour characters the single and double quantities both *show* that colour in their plumage. With the Recessive colour characters it is only the *Double* quantity that shows that colour with the Single quantity carrying it in hidden (split) form.

SEX-LINKED INHERITANCE

The Sex-linked colour characters which are Recessive and inherited differently from the ordinary Recessive kinds as they are carried on the X chromosome pair which controls the sex of the birds. The cock's sex chromosome pair is described as XX and can carry colour characters on one or both members of the pair. To show visually a Recessive colour character it must be present on both members of the pair as if only on one the bird will be a carrier or split. As a hen bird has only one X its partner a Y is a sex determinant and it must show visually any recessive colour character carried on the single X. The Recessive Sex-linked colour characters operate quite independently to the other characters that are carried on the ordinary chromosome pairs. The rules governing Recessive Sex-linked inheritance are as follows:

Double colour character (cock) paired to single colour character hen gives cocks and hens *all* showing visually that colour character. An example: Fawn cock to Fawn hen, gives all young with the Fawn character.

Double colour character (cock) paired to Normal hen gives all single colour character carrier cocks and all single colour character (visual colour) hens. Example: Fawn cock to Normal Grey hen gives Normal carrying Fawn cocks and Fawn hens.

Normal Grey cock paired to single colour character (visual colour) hen gives all Normal Fawn carrying cocks and all Normal hens.

Normal (single colour character and paired to single colour character (visual colour) hen gives Normal/single colour character cocks, Normal hens, single colour character (visual colour) hens and double colour character cocks, there is no visual colour difference between the split and the pure cocks.

Normal/Single colour character cock paired to Normal hen gives Normal cocks, Normal/single colour character cocks, Normal hens and single colour character (visual colour) hens.

The varieties covered by the above set of rules:

Dominant	**Recessive**	**Sex-linked**
Normal Grey	White	Fawn
Dominant Dilute (Silver and Cream)	Recessive Dilute (Silver and Cream)	Chestnut-flanked White
Crested	Pied	Light-back
	Penguin	Albino
	Black-breasted	
	Yellow-beak	

By combining two of more of the above colours many further forms can be created by the keen breeder. These are not new colours but a combination of already existing colours. As an example the crossing together of the Recessive Pied, Sex-linked Fawn and the Dominant Crested characters, the Crested Fawn Pied will be created. From this example it will be seen there is great scope for breeding quite spectacular birds.

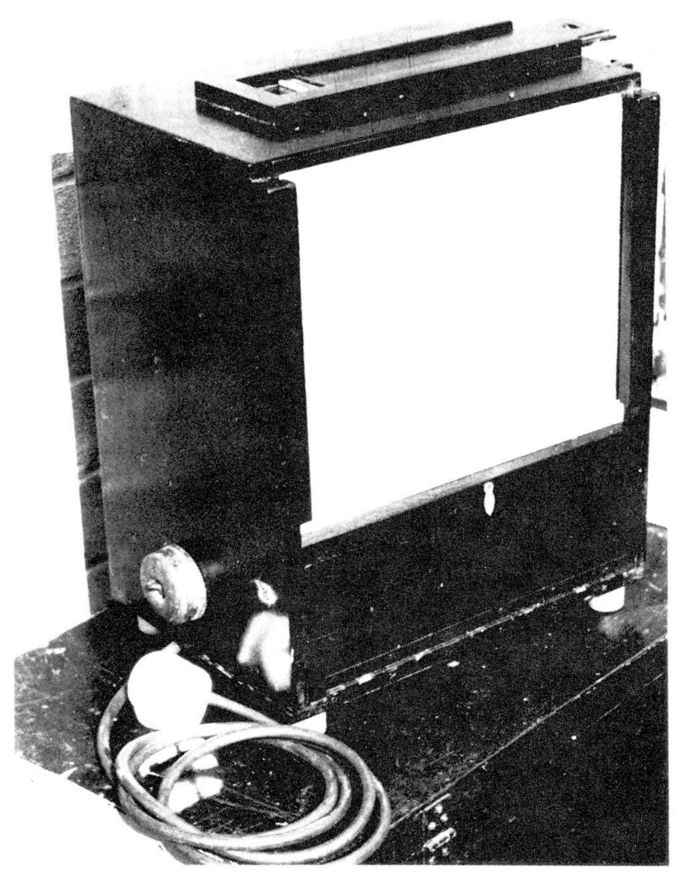

A typical hospital cage used when cage birds are 'off colour'

5
THE ZEBRA FINCH SOCIETY

A typical Bird Room — note feeders and other essentials

Chapter 5

THE ZEBRA FINCH SOCIETY

By now Zebra Finches were rapidly becoming a recognised Exhibition species and classes for several of the colour varieties were being scheduled at more and more of the larger shows. It was in the year 1952 that The Zebra Finch Society was launched after a meeting held in Birmingham by enthusiastic breeders. Some ninety Zebra Finch breeders joined this new Society during the first year and a few of these are still active members.

THE FIRST OFFICERS

The first officers of the Society were President: Allen Silver, F.Z.S., Vice President: P.A. Birch, Chairman: A.J. Wilkins, Hon.Secretary: S. Moulson, Hon.Treasurer: F.G. Cannon, with an Executive committee consisting of Miss J.P. Uncle, Messrs. P.A. Pope, E.J. Hounslow, A.J. Rooke, G.B. Shaw and T. Maughan. Rules were formulated, standard and colour classifications were developed, patronage given to Open Shows and Messrs. A.C. Hughes were appointed official ring makers.

EARLY SHOWS

Some six shows were given Patronage with the very first one being held in the Town Hall, Hove, by Brighton & Hove C.B.S., and twenty four entries were received. The largest entry of that year was at the National Exhibition where forty three pairs of several colours were benched. In 1953 the first Year Book was issued with S. Moulson acting as Editor in addition to his other offices. Having had and bred Zebra Finches for many years, my wife and I joined the Zebra Finch Society and the following year I took over the office of Editor which I held until 1973. In 1962 S. Moulson retired from the joint office of Secretary and became Hon.Treasurer and J.A.W. Prior became Hon.Secretary with Mrs. H. Rogers as his Assistant Secretary.

GROWTH OF THE SOCIETY

Since the late 1950s and early 1960s, the Society has steadily gained strength and prestige in the cage bird fancy and its rulings are recognised throughout the world. An important progression of the Zebra Finch was when in 1957 the Zebra Finch Society declared that the Normal Zebra Finch and all its colour mutations to be truly domesticated birds and would no longer be classed as a Foreign species. This declaration was enthusiastically received universally and the Zebra Finch is now recognised as being fully domesticated along with Budgerigars, Canaries and Bengalese. I think that the following extracts from an article by J.A.W. Prior, Hon. General Secretary in the Society's Silver Jubilee book of 1977 gives readers an excellent resume of the progress of the Zebra Finch. **"Since the formation in 1952 the Zebra Finch Society has grown to become in its Jubilee Year one of the largest international specialist societies in the bird fancy."**

SIZE OF FANCY

Zebra Finches are traditionally regarded by most bird fanciers as an 'easy starter' for those entering the Foreign Bird Fancy, but over the years many fanciers have become devoted to the Zebra Finch in its own right recognising its true value and the challenge it offers in both mutation breeding and exhibition.

Membership reached 1151 in 1976 and the figures for the Jubilee Year foretell the same steady increase experienced over the last fifteen or so years. It is an interesting fact, however, that the Society has in its twenty years existence had 5247 members pass through its books.

There is also the undoubted hidden population of those who have in the past, or do so now, keep Zebra Finches but have never entered membership of the Society. A Society census at the National Show in 1974 revealed that only one in three Zebra Finch owners and breeders belonged to the Society. The two-thirds majority recognise it only to be a nice friendly colourful little bird that they keep in their aviary . . .

On this basis and noting the fact that half of The Zebra Finch Society members only purchase rings (23,500 in 1976) well over 100,000 Zebra Finches must be bred in the United Kingdom alone each year. . .

The Zebra Finch Society is of course supported by Area Societies in Scotland, Scottish and Northern Counties, Yorkshire, Midlands, East Anglia, West of England and Kent each of which accepts and maintains the parent Society's rules, show classifications and *standards.*

Zebra Finch Societies also exist in the United States of America, Canada, Australia, New Zealand, France, Italy, Denmark, West Germany, Belgium, Sweden, and are all affiliated to The Zebra Finch Society.

The membership of The Zebra Finch Society is still steadily growing and in the year of 1983 had reached approximately 1500. The limited space needed to keep Zebra Finches, the many differing colour mutations and their combinations and the consistent way they reproduce are the factors that are encouraging fanciers to take up this species. Many hundreds of Show Promoting Cage Bird Societies have been given one of the categories of The Zebra Finch Society's Patronage and from these one was selected to hold the Club Show. This was so until 1974 when the first Zebra Finch Society independent Club Show was arranged and staged at Urmston, Manchester, and attracted some three hundred and seventy six entries. However, it was at Oxford in 1980 that the magnificent entry of four hundred and forty-five entries were received, this beating the record of 1974.

It is interesting to note that the Best in Show and Best Champion Pair were Fawn breeders and shown by M. S. Wren and Best Novice were also Fawn Breeders and benched by Mr & Mrs R.A. Clarke. This points to the popularity of Fawns and that young birds are frequently of better colour, type and quality than adults which can become coarse and over fat with age. Since the record Oxford Club Show the entries of subsequent Club Shows have been quite good, but have not reached that record total.

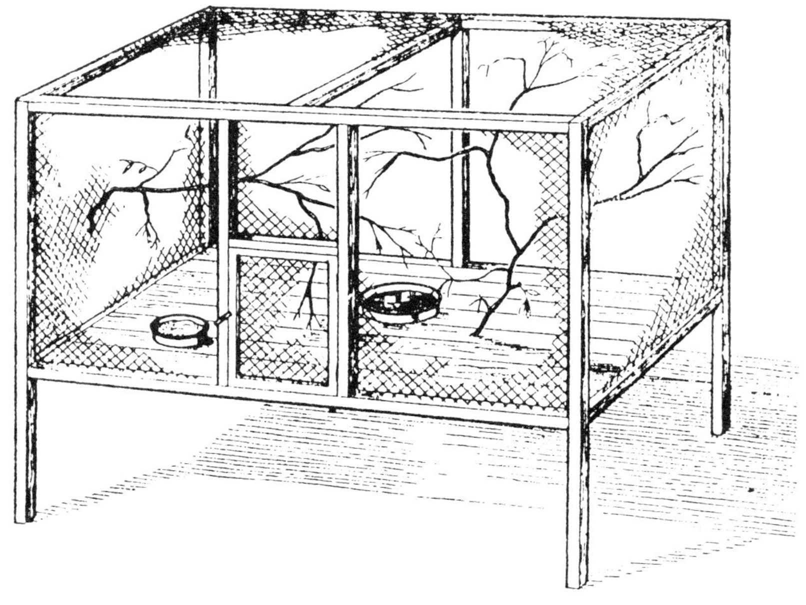

A simple indoor flight which gives more space than a cage.

PART 2

MANAGEMENT

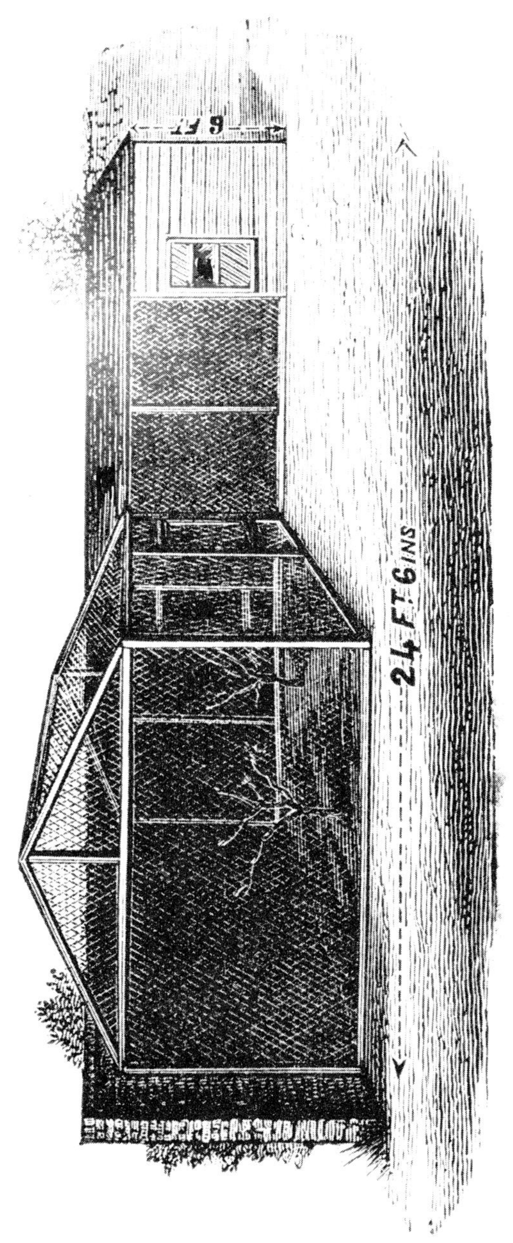

A Large Outdoor Aviary

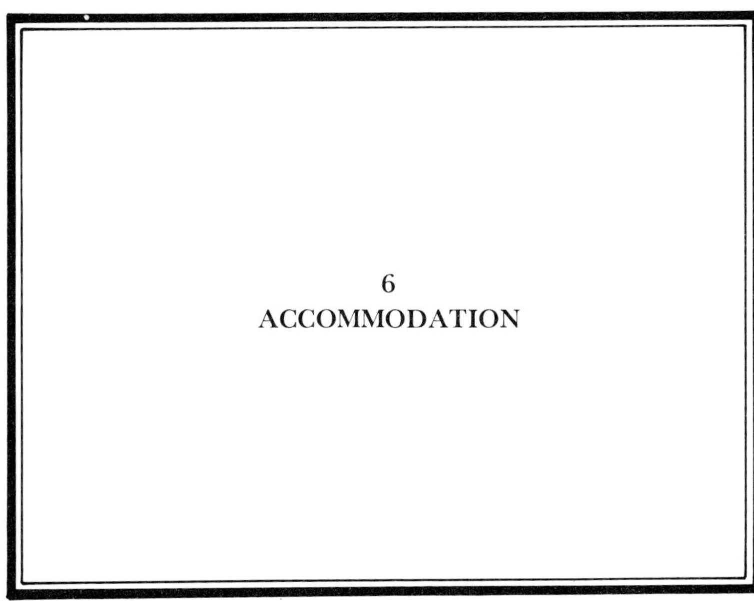

6
ACCOMMODATION

ON THE AVIARY

" There is no necessity . . . that such a building should be anything but attractive in appearance and at the same time maintain its efficiency as a home for birds"

L. P. Luke *A Pioneer Aviculturist*

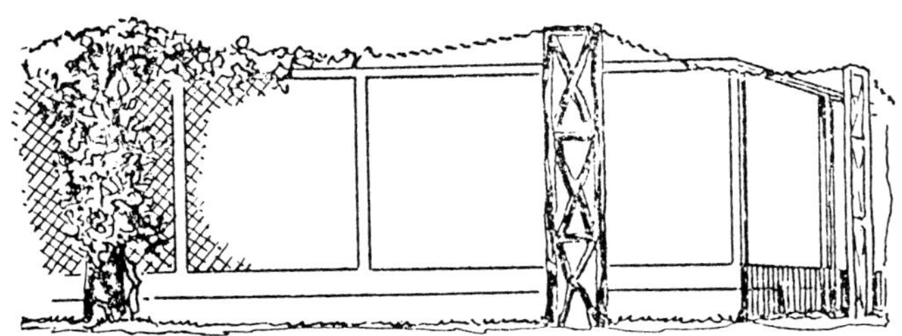

Rambler roses, trained up rustic poles or lattice, and along ropes or chains, make effective screens

Even a simple aviary looks impressive when mounted on a rockery

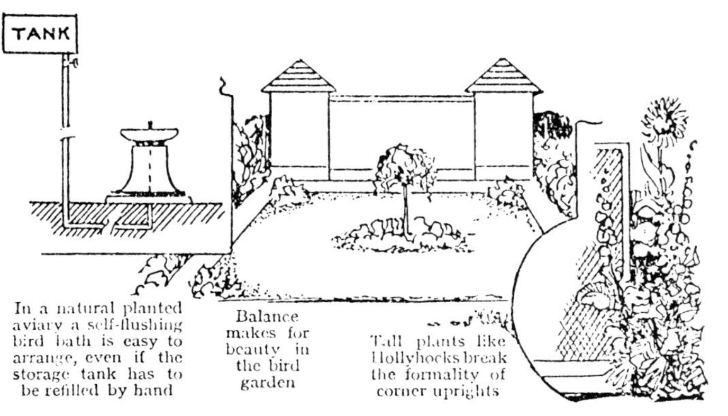

In a natural planted aviary a self-flushing bird bath is easy to arrange, even if the storage tank has to be refilled by hand

Balance makes for beauty in the bird garden

Tall plants like Hollyhocks break the formality of corner uprights

The Bird Garden can be very attractive

Chapter 6

ACCOMMODATION

ALLOW SUFFICIENT SPACE

After having read about the development of Zebra Finches contained in the first part of this book I feel that many people will want to try breeding these delightful little domesticated birds. Once having decided to take up the keeping and breeding of Zebra Finches the first thing that has to be considered is the adaption of acquisition of suitable accommodation. Fortunately Zebra Finches are easy birds to house and can be kept and bred in many kinds of structures ranging from cages to flighted aviaries. Although they are quite small birds about 4"–4¼" overall length, they are extremely active and require a reasonable amount of flying space if they are to be maintained in a good healthy condition.

CAGES

A great many collections of Zebra Finches are housed in **cages** and these in my opinion should not be less than 30 to 36 inches* in length, some 15 to 18 inches in height and 12 to 15inches deep. Cages of these dimensions will give reasonable flying space for a single breeding pair with an inside nesting box or some eight non-breeding birds. Perches should be kept to a minimum so that the greatest distance

***Multiply by 2.54cm to convert to metric measures – see Standard Cage Measurements in The Show World chapter.**

between them can be achieved. There are suitable ready-made cage fronts with doors large enough to allow a nest box to be put in and taken out with ease. The seed, water and grit vessels should be placed clear of the perches and it is also beneficial if **flomatic** type drinkers are also provided. I like to use washed sand as a cage floor covering, but coarse pine sawdust is also used quite successfully by many Fanciers.

CONSTRUCTION

Cages themselves can be constructed of plywood, planed light wooden boards, hardboard or a combination of these materials. It is best if the cages can be built in blocks so that at non-breeding times two or more can be made into useful flight cages by removing the dividing slides. It is up to each individual breeder's preference whether sand trays are fitted, but I would suggest that good *deep* front rails are provided to help to prevent the undue scattering of seeds by the inmates.

CONTAINERS

Grit, water and seed containers should be placed clear of perches and nest boxes. These vessels can be of any material that is easily cleaned, but for water they should not be of the porous kind. Bird accessory stores usually have a good selection of different sized dishes from which the Fancier can choose those of suitable size for his or her particular requirements.

PERCHES

The feet and legs of all cage birds can soon become stiff or sore if the perches used are not of different thicknesses which will give the birds a change of grip thereby exercising the muscles in feet and legs. Various materials can be used for perching such as machined dowelling or natural wooden branches with or without bark; smooth cane however is unsuitable as it can cause birds to slip when mating. The decoration both on the inside and outside of the cages can easily be done with distemper, paint, or emulsion paint, with the latter I think being the most suitable for their purpose.

SMALL PENS

Small pens with or without flights are excellent for breeding under partial control with two or possibly three pairs allocated to each pen according to its size. It is important that each pair has ample space for their nesting area as they are very protective of their nesting sites and will drive away intruders. They are actually sociable birds and in the wild several pairs will build their nests in the same bush with each respecting the others immediate territory. Pens can vary in their size according to the space available to the breeder and a good size is I think some 27 inches wide, 4 foot deep and 6 foot 6 inches high. If outside flights are added these can be of similar size to the pens themselves and are best made of ½" mesh wiring. During the non-breeding periods these pens can hold some twelve to fifteen birds making them ideal for stock accommodation. The seed, grit and water vessels can be similar to those as used in the cages. Coarse washed sand is ideal for floor covering of any flight and the pen floors can be covered with the same material, coarse pine sawdust or a mixture of both.

OUTSIDE FLIGHTS

It will seem that Zebra Finches will breed more freely in outside flighted aviaries than in other kind of housing and this is only natural with a gregarious species. Aviaries can be of any size or shape and other existing buildings can be adapted to make good flighted aviaries. The drawback with flighted aviaries is that they take up quite a large area of ground and gardens are on the small side nowadays and of course there is no real control of the breeding pairs. However, this latter can be solved to some extent if each of the desired pair are mated and the pairs kept separately for a week or so before releasing them into the aviary. Again, the perches, seed, grit and water vessels can be similar to those as used for cages but of a larger size for the vessels and the floor covering can be the same.

LANDSCAPING

Undoubtedly a planted flight will add an attractive addition to any garden aviary and will give the breeder ample scope for individual design. Here the floor can be covered with grasses and shrubs, bushes and small trees can be used as perching and decoration much to the satisfaction of the inmates. Unlike the Parrot-like species Zebra Finches will not gnaw the growing plants and quite a wide range of varieties can be included in the planning. One of the fastest growing climbers is the Russian Vine which soon makes excellent cover for the birds. The various varieties of decorative Ivies are useful for covering bare areas and Lonicera nitida, Lavenders, Privet and Conifers together with many other kinds of decorative shrubs (not the poisonous fruit kinds) all help to make the aviary pleasing to look at and always an attractive feature. With a planted flight the breeding pairs will often completely make their own nests, but nest boxes should always be provided as should some nest building materials.

PROTECT AGAINST VERMIN

Both the flight and sleeping quarters should be wired against intrusion of vermin which if not prevented can soon ruin the structure and harm the birds. Pieces of small mesh wire netting, L shaped, should be sunk in the ground completely round the flight and sleeping house. Such a precaution will keep out the vast majority of mice and rats, stoats and weasels, but it is advisable to inspect the wiring periodically just in case a break has occurred. Similarly all woodwork should be frequently examined, particularly near the eaves and many blind spots. With these precautions it will help to keep the aviary in a good sound condition and the birds all safe.

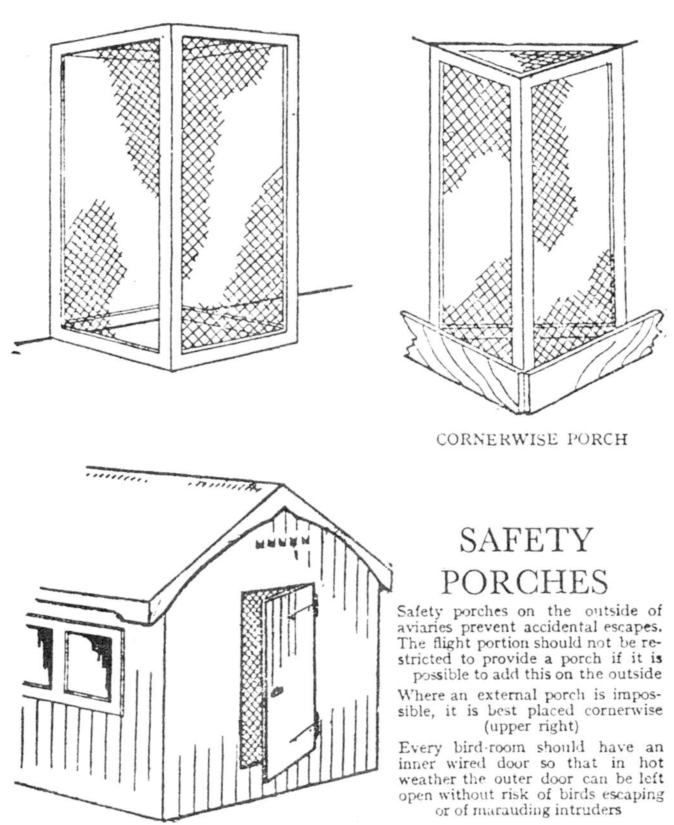

Safety Porches are a desirable feature

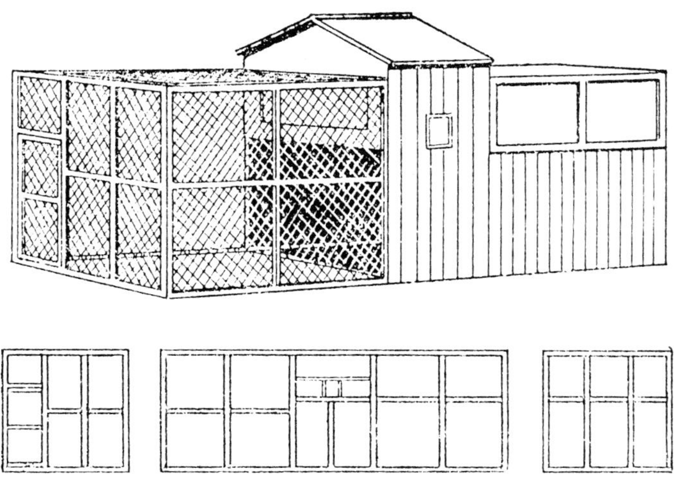

Basic arrangements of suitable structures. That shown below is an ingenious type, without shelter shed, designed by Hylton Blythe for British birds

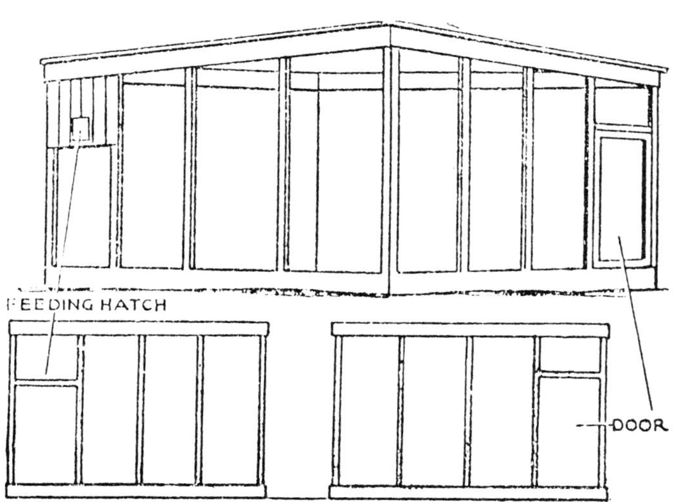

The Summer Aviary (designed by Hylton Blythe)

7
FOODS AND FEEDING

Different types of food container – half-moon grit pots, finger drawer and various jars

Chapter 7

FOODS AND FEEDING

AIM FOR A VARIED DIET

The diet of Zebra Finches is quite a simple one and their foods are easily obtainable in all areas which make their keeping and breeding attractive to first-time bird keepers. Whether they are housed in cages, pens or aviaries, they all eat the same kind of seed mixtures. The bulk of the seed mixtures consist of a small yellow millet known as Panicum millet and they will exist quite well on this seed alone. It is nevertheless advisable that all cage and aviary birds should be given a varied diet and this of course includes Zebra Finches. Small quantities of white millet, large yellow millet, Japanese millet and small canary seed should be mixed with the Panicum millet. In cold weather a pinch or two of Niger, an oil-bearing seed should be added; it must only be used sparingly as it can cause fatness if eaten in any quantity. As most breeders have their own ideas as to the quantities of each seed used in a mixture it is difficult to give exact amounts.

TRY TO ESTABLISH INDIVIDUAL TASTES

Individual Zebra Finches have their own taste for seed and in an aviary all the mixed seeds are eaten by the colony. In cages it will soon be discovered which particular seeds they discard and a mixture can be made up to suit their particular requirements. All Zebra Finches are very partial to millet in the spray and sprays either dry or soaked are a useful addition to their diet when feeding young and some pairs would rear entirely on sprays if given the opportunity. However, mixed seeds should always be available to the birds so that when the young leave the nest they have an opportunity to find their regular diet. In addition to the seeds mentioned above many birds will find interesting and beneficial seeds if given dishes of mixed wild seeds to pick over. This also helps to keep both old and young birds thereby helping to prevent undue squabbling.

SOAKING SEEDS

Mixed seeds soaked in clean cold water for some twenty-four hours then drained are a welcome change for feeding parent birds. If on the other hand seeds are soaked and then allowed to sprout they will serve as an excellent green substitute when other kinds are difficult to obtain.

SOFT FOODS

Most but not all Zebra Finches will appreciate some kind of soft food if given occasionally all through the year, with the quantity increased when young are in the nest. Small cubes of moistened wholemeal bread are eaten readily by the majority of birds. Too much soft food should not be offered as it can make the droppings very loose and will form a hard crust around the nest when full clutches are being reared and can result in foot deformities. There are numerous excellent patent soft foods on the market and I would advise breeders to try several to find the one that is most favoured by their particular stock of birds. Soft foods must be mixed and given fresh daily and any that is uneaten removed as stale food can quickly upset the stomachs of both old and young stock and particularly so in hot weather periods.

FRESH GREEN FOODS

Fresh green foods are another very important item of diet

for Zebra Finches. Although all birds do not like every kind they will eat most varieties with their favourites being chickweed and seeding meadow grasses. These greens can be varied with lettuce, heart of cabbage, plantain heads, young dandelion leaves, spinach, shepherd's purse, sowthistle and my own birds have a great liking for the pith inside of marrow. Here again all green should not be allowed to go stale but removed at the end of each day. During the breeding time green food can be a little problem if not given tied in small bunches or firmly fixed in a clip, as some birds will use it as nesting material even when the clutches of eggs are already being incubated. This making of 'sandwich' nests can cause clutches of eggs to be spoiled and therefore all surplus nesting materials should be removed as soon as the first eggs appear.

GRITS REQUIRED

As well as their foods the birds must always have access to plenty of various grits, cuttlefish bone and mineral blocks. Grit is an essential item to all birds and Zebra Finches seem to get through quite a lot of this commodity. A further excellent source of calcium is dried crushed shells from domestic hens' eggs which the birds will eat regularly. The egg shells should be thoroughly dried by baking slowly in an oven, then crushed finely with a rolling pin or some such article and stored in an air-tight container.

HEAT AND LIGHT

Good food, clean water and plenty of exercise will keep Zebra Finches in a fit healthy condition without any artificial heating. Although artificial light is not necessary it can be very useful if it can be laid on to the bird room or aviary for giving the birds a late feed in the winter months or preparing birds for exhibition.

A Nest Box which can be made quite easily

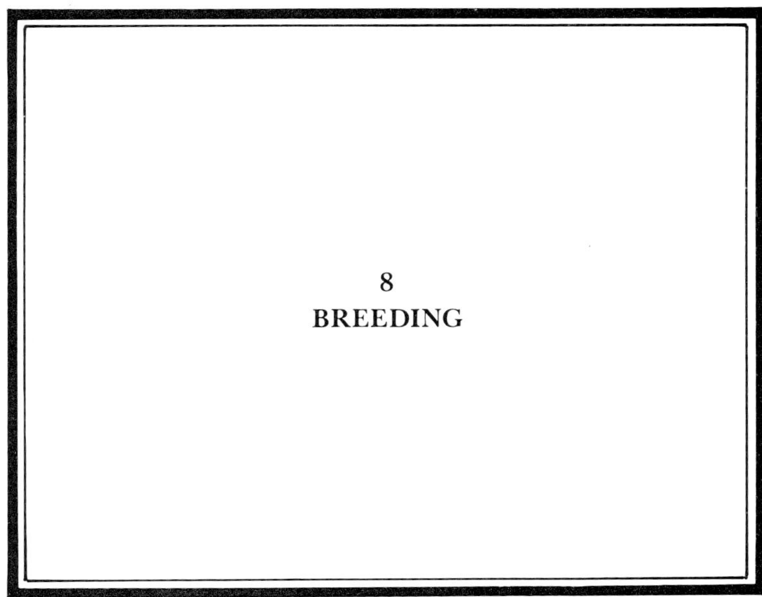

8
BREEDING

ESSENTIAL FACTS

Laying: four to seven eggs. *Incubation:* eleven days. In the daytime the sitting is alternate, at night both consorts together. The young are fed for a considerable time, even after their flight they like to sit with their parents in the nest.

Dr. Arthur G. Butler

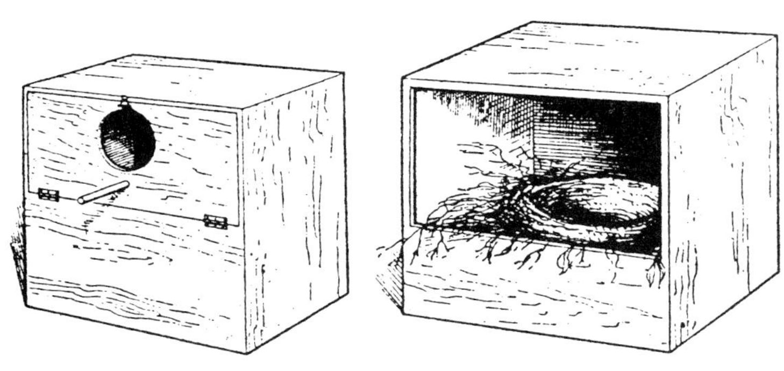

Typical nest boxes – closed type (left) and open-fronted (approx 6" x 6" x 6" – 150 x 150 x 150mm)

Chapter 8

BREEDING

DECISION TIME

When the bird house or aviary has been constructed and found satisfactory and a stock of fit healthy birds have been purchased, Fanciers thoughts invariably turn to the planning of the breeding programme. The commencement of this very important operation is of course controlled by the time of year as breeding should only be attempted in the early part of the year when the weather conditions are suitable for successful rearing of healthy young birds. It is only when both cocks and hens are in full breeding condition and are not less than ten to twelve months old that they should be allowed to breed.

BREEDING CONDITION ESSENTIAL

If one or both members of pairs are not in condition they may attempt to nest and give unsatisfactory results. When in full breeding condition, birds will appear very active, inclined to chase other birds and their eyes should be bright but it does not matter if a few feathers are broken or missing. Fit cock birds will stretch up their necks and utter their strange unmusical courtship song, often with a piece of grass, twig or feather in their beaks. Hen birds will similarly be active, bright eyed and forever searching for possible nesting sites, carrying around any material that could be used for nest building. Should cock birds be within sight they will call vigorously, answering the cock birds courtship songs. When both members of selected breeding pairs are seen to be quite ready they can be allowed to go to nest providing of course the weather conditions at that time are suitable.

STARTING TIME

Experienced Zebra Finch fanciers have their own ideas just when to start breeding each year and this is often controlled by the part of the country in which they live. Those living in the northern counties usually start a week or two after those in the South and South-West. For first time breeders I suggest that the end of February to the beginning of March is a reasonable time to make a start again, providing the weather is set for fair and the breeding pairs are fit and housed in good dry, draught-proof quarters. If breeding is started too early there is always the possibility of hens becoming egg-bound, eggs or young getting chilled, and the chicks dying. Should at any time a hen or a cock be seen to be fluffed up and looking sorry for itself it should at once be caged and taken into a warm, even temperature. Sick birds should be given their usual seed mixture, millet sprays and clean water and usually after a few days in the warm eggs will be laid and the sick birds recover. Birds that have been taken into the warm should not be returned to their usual quarters until the breeder is sure they are fully fit again, and the weather is reasonable if put outdoors.

NEST BOXES

The nest boxes should be hung in place after the pairs have been in their breeding quarters for a few days so they have had time to settle in their new accommodation. Where there is more than one pair in a compartment it is best to allow two nest boxes per pair so that the birds have a choice and when they have made their selection the unused ones can be removed and used as spares. The nest boxes should be 5 inches cube in size with either a half open front or a front entrance hole of just over an inch in diameter with a hinged top for inspection. Nest *baskets* can be used in aviaries, but they do pose a problem for inspecting eggs and the ringing of chicks.

I usually half fill the nest boxes with soft dried grasses and then give the birds a further supply so they can complete the building of the nests. They also appreciate a few soft feathers or a little cow hair with which to line the nests, but horse hair or similar tough fibrous materials must be avoided to prevent the birds getting tangled. With single pairs in breeding cages only one nest box is needed and this is best hung facing the light. If a piece of cardboard is fixed to the wire front opposite the box, this will give the birds a little privacy and make them feel more secure. Once the first eggs have started to be laid, all unused nesting material should be removed thus preventing the building of 'sandwich' nests which can soon spoil the clutches.

EGGS

The number of eggs per clutch can vary from six or more, with the average being five. Incubation may start after the second or third egg has been laid, this means that in the clutch of five the last eggs will hatch a few days after the first. The majority of Zebra Finches are most attentive parents and will feed and look after all the chicks in a nest with great care.

The actual incubation period is eleven to twelve days for each egg, according to the time of year, with both parents sharing the duty. Young Zebra Finches develop very quickly and are mostly ready to leave their nest boxes when sixteen to eighteen days old.

Their parents will continue to feed their broods some fourteen days or so after they have flown and they can be removed from the care of their parents as soon as they are seen to be eating well on their own. It is best to remove the young as soon as possible so that the parent birds can be free to proceed with a further clutch and not be worried by the previous batch of chicks.

CLOSE RINGING

Most breeders like to close ring their young birds as a permanent method of identification and it is a must when they are being exhibited in Breeders Classes at any Zebra Finch Society Patronage Shows. These closed, coded, year dated coloured metal rings can be obtained through membership of **The Zebra Finch Society**, Hon Secretary, **J.A.W. Prior, 87, Winn Road, Lee, London S.E.12 9EY.** Young Zebra Finch chicks are very small and, therefore, when they are being close ringed they should be handled with great care so as not to damage their growing feet and legs.

Because like all baby birds they develop at different rates, it is difficult to state an exact time when the rings should be put on the youngsters. Some breeders like to ring their chicks when they are very small, but others like myself prefer to wait until the feathers are starting to sprout and the birds are more easy to handle and the rings less likely to slip out of place.

RINGING PROCEDURE

The rings should first be slipped over the three front toes, then along the shank with the hind toe, which is then pulled through with a sharpened matchstick leaving the ring in position on the shank. If it is possible, new breeders should try and see how an experienced breeder rings his or her birds as this will inspire confidence when ringing their own chicks. In addition to the closed rings there are the split metal and plastic kinds for marking birds that have not been closed ringed. The split plastic rings are very useful for marking families of birds so they can be identified without having to catch them up to read their closed ring numbers.

KEEPING RECORDS

Whether the breeder is interested in exhibiting, colour breeding or both it is most essential that careful records of all birds are kept in some detail. To keep an accurate breeding register all stock used for breeding must be ringed preferably with closed coloured coded year dated rings. As the chicks are ringed their numbers should be entered into the register against the appropriate pair and as soon as the colour can be seen this should also be included as should the dead in shell, addled, or clear eggs. If any chicks are moved for some reason into another nest before they are ringed, they should be carefully marked with a felt pen and when old enough ringed and the numbers entered under their true parents.

THE ZEBRA FINCH SOCIETY

When taking up the hobby of keeping and breeding Zebra Finches one of the first things that should be done is to join **The Zebra Finch Society** or one of its affiliated area Societies, thereby getting the many benefits of membership. It is also a good idea to join a local Cage Bird Society so that members keeping Zebra Finches can be contacted and much useful knowledge gained. To keep in touch with the up-to-date matters in the Avicultural world, that excellent weekly paper *Cage and Aviary Birds* should be taken. This paper also gives breeders an opportunity to buy fresh stock and to sell their own surplus birds.

Alternative nest pans (3 types)

A small lean-to Aviary

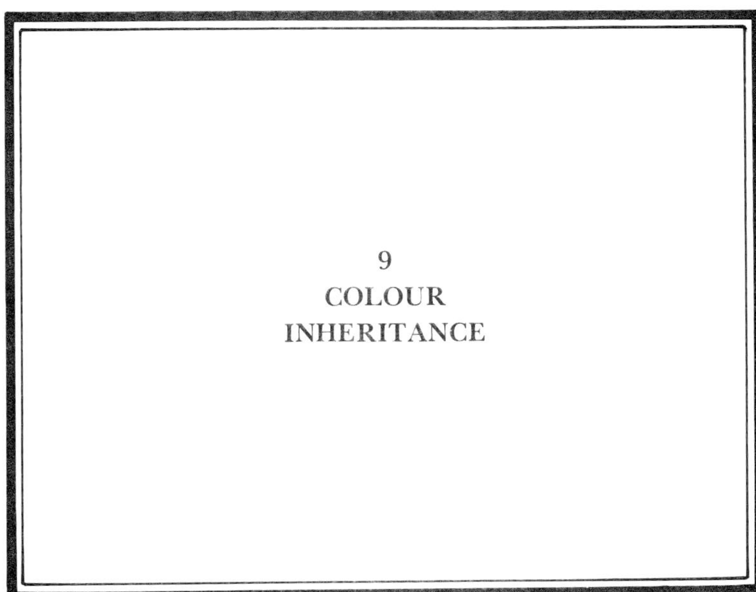

9 COLOUR INHERITANCE

Chapter 9

COLOUR INHERITANCE

MANY COLOURS AVAILABLE

When breeders first come to the Zebra Finch Fancy they are often somewhat bewildered by the quite large number of different coloured varieties that are now available. It is always difficult to say just what colour or colours the beginner should have for starting up an aviary. The easiest colours to obtain in most areas are the Normal Greys and Normal Fawns and after having some of these for a season or two to get the feel of breeding Zebra Finches, the breeder may wish to add to the stud some of the other colour varieties as discussed in the first part of this book.

MUTATIONS

As it will have been seen, the original colour of the Normal Grey or Wild Type and from them all other colours and combinations of colours are mutations which have occurred and been established over the years. These mutations are divided into three breeding groups – Dominants, Recessives and Sex-linked Recessives each being inherited in its own particular manner.

DOMINANTS

The Dominant Group consists at the moment of Dominant Silvers (Creams) and the Crested. The Recessives (explained below) form a larger Group and are Whites, Penguins, Pieds, Recessive Silvers (Creams), Black-breasted and Yellow-beaked, and the Sex-linked Recessive kinds are Fawns, Chestnut-flanked Whites, Light-backed and Albinos. The latter kind are not yet to be had in this country, although they are breeding quite freely in Australia. There are of course many combinations of these character groups resulting in a surprising number of various mixed shades and patterns.

The Dominant Silvers which can be found in either single or double quantities of the Silver character both give the same visual colour result. Dominant Silvers are in fact Greys with the addition of the Silver Character which alters their colour to a Silver-grey colour. If the same Dominant character is incorporated with Fawn the rather pleasingly coloured Creams will result.

There can be a Dominant Silver (Cream) form of all the other varieties although in some cases the combination does not result in very attractive birds.

The **Crest** is the Dominant character and again can be had in all varieties. It will be seen that birds bred from one Crested parent but do not have a crest, i.e., Crestbreds, help to produce a greater number of better Crested birds when paired to Crests than do straight Normals.

Dominant mutations can quickly be increased as when say a Dominant Silver (single character) is paired to a Normal Grey, half the young will be Grey and the other half Silver (single character). When a Dominant Silver (double character) is paired to a Normal Grey all the resulting young will be Silver single character.

RECESSIVES

With the Recessive varieties such as White when paired to Grey, all the resulting young are grey in colour but carry the character for White in their genetical make-up and are known as Grey/Whites. If two different kinds of Recessives are paired together all their offspring will be grey in colour but carry both Recessive characters in their genetical make-up. As an example, Pied Grey to Penguin Grey will produce all Grey/Penguin Pied which when mated together will give the following colours – Grey, Penguin Grey, Pied Grey and Penguin Pied. This will indicate to the breeder that colour characters can be transferred on to other colours, giving composite birds that are very often exceedingly intriguing from a colour angle.

SEX LINKAGE

I now turn to the Recessive Sex-linked kinds where the sex of the birds used has a big influence on colour. The reason for this is quite simple as the pair of Chromosomes that control the sex of the birds differ and those for the cocks are written as **XX** and those for the hens as **XY** and it is only the **X** that carries colour characters with the **Y** directing the sex. So with a hen which has only one **X** that can carry colour characters it means that the colour carried by the **X** must show in the plumage. For example, take the popular Fawn where a Fawn cock paired to any non-Fawn hen will produce all cocks split for Fawn and all visual Fawn hens. Fawns can be identified as soon as they hatch by the pinkish colour of their eyes as opposed to the black colouring of the non-Fawn varieties. When a Fawn hen is paired to a non-Fawn cock all their young are normal in colour with the cocks only being split for Fawn. The same pattern is followed by all the different sex-linked kinds.

An interesting member of the Sex-linked Group is the Light-back which has an affinity with the Chestnut-flanked White character. If a Light-back is paired to a Chestnut-flanked White all the young will be Light-back split Chestnut-flanked White, i.e., single character Light-backs. When a Light-back Chestnut-flanked White is paired to a Chestnut-flanked White they will produce Light-back/ Chestnut-flanked White and Chestnut-flanked White cocks and hens in equal numbers. Chestnut-flanked Whites on the other hand *cannot* be split for Light-backs. From the above paragraphs breeders will have seen how different colour characters are handed down from one generation to another.

A Black Cock from a drawing by C. Harding

10 THE SHOW WORLD

A Young Fancier

Chapter 10

THE SHOW WORLD

EXHIBITING

After having kept and bred Zebra Finches for a few seasons many Fanciers think about exhibiting some of their birds particularly so the ones they have actually bred. Exhibiting is another aspect of aviculture which helps the breeder to improve the standard of the stock and promotes a healthy spirit of competition amongst members of local and specialist Societies.

PREPARATION

Breeders will, I fee sure, quickly realise that exhibiting Zebra Finches requires more than just taking birds from an aviary or stock cage, putting them in show cages and taking them to a show. To get the best out of the birds they must have some **training** as how to behave when in show cages. If birds are put into show cages for half to one hour several times a week they will quickly settle down and take to show cage life quite readily. Not only should they be caged but they should be moved around in the bird room so as to simulate the movement at shows.

If a show cage is hung over an open door of a stock cage the birds can be encouraged to go in and out by putting a millet spray or some green in the show cage. This procedure soon gets the birds accustomed to show cages and when shut in a cage they will settle down quite well. Fitness, clean perfect condition of all feathers, steadiness, good colour and type are all essential ingredients that go to make good exhibition birds.

MATCHED PAIRS

Zebra Finches are always shown in matched pairs of the same colour variety and need not be breeding pairs as often birds from the same parents will make good matched exhibition pairs. Each pair should be selected for its exhibition attributes and should be of the same shape, colour and size even if a single aspect has to be sacrificed in the cause of uniformity. Exceptional birds should **not** be matched with nondescript partners; it is far better to have a reasonably good but balanced pair, than say an exceptional cock and a poor hen or vice versa. Exhibition birds should be steady in their cages when lifted by the judge to examine them closely as birds that flutter about cannot be properly assessed no matter how good they may seem at home.

STANDARD

The **Zebra Finch Society** has laid down certain *standards* for exhibition birds as indicated in the following paragraphs and exhibitors should follow these at all times as near as possible. They are as follows:

1. **Condition.** Birds should not receive any award unless in perfect show condition. Missing, ragged or soiled feathers and missing claws or toes constitute show faults.

2. **Type.** Bold throughout and of the 'cobby' type, giving the birds a look of substance, wings should be carried evenly to the root of tail.

3. **Markings.** (cocks) Chest bar distinct and clear cut, not less than 1/8″ wide and of even width throughout. Side flankings should be prominent, extending from wing butts to end of rump and decorated with round clearly defined white spots. Beak coral red with feet and legs

deep pink. (the colour of feet and legs will vary with certain varieties). All markings where applicable to be clear and distinct. Hens as for Cocks less cheek patches, chest bar and side flankings, beak a paler shade of red. Male markings on hens are definite show faults.

It should be noted that colour feeding of Zebra Finches is debarred as also is the trimming of wings and/or tail.

The Zebra Finch Society Show Cage

MEASUREMENTS

The specifications and dimensions for the standard show cage are as shown below and opposite.

Wood. Top, sides, bottom, false roof and front rail, 6mm thick. Back 4mm thick plywood.

Perches. Overall length 100mm. 10mm diameter. Rear boss, 25mm Diameter projecting 10mm.

Front Rail. Right hand edge of the drinker door cut at 45°. Escape wire (painted) white fitted into other end. 'S' hook fastened to centre of drinker door with a staple. Drinker door fastened by a brass desk turn painted black. Door hinged using pin through front rail drinker door. Zinc clip screwed to the inside of door to carry white plastic drinker.

Wire front. Comprising twenty-three wires, 12.7mm centre to centre. Double punched bar at top set 5mm apart, for fixing two wires left top and bottom painted black.

Colour. Inside white. Outside black.

Floor Covering. Any suitable seed.

No Maker's name to appear on cage.

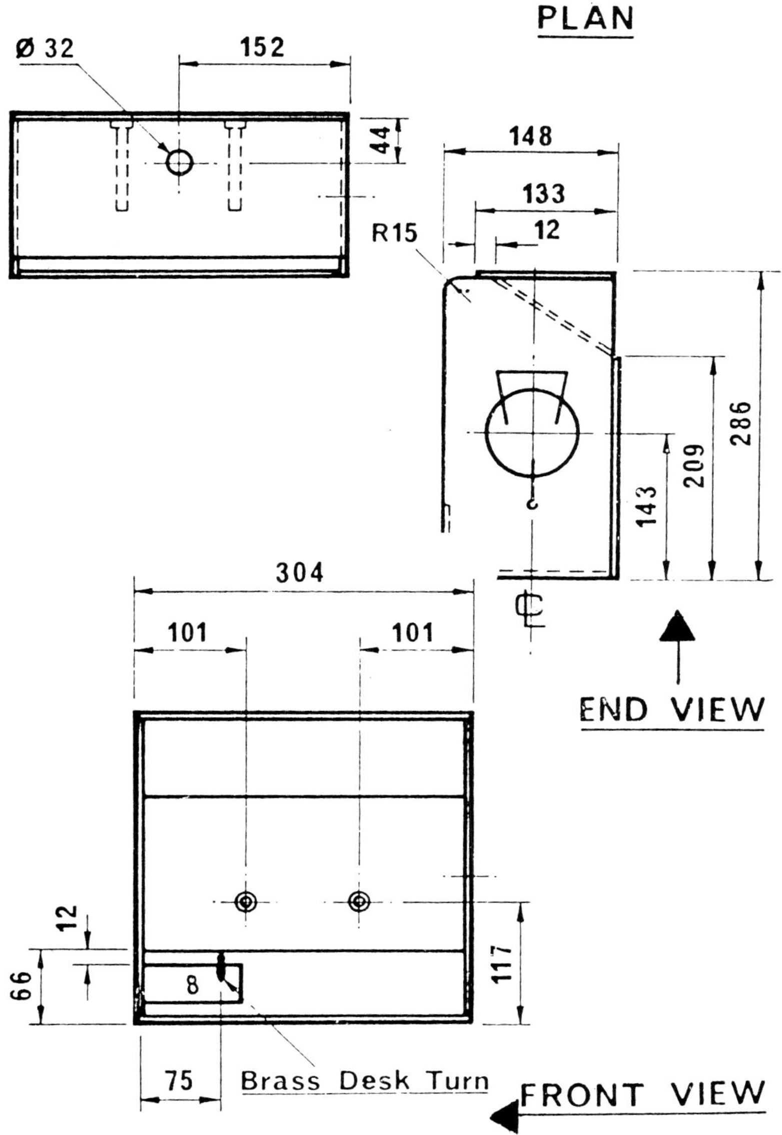
DIMENSIONS In Millimetres
PLAN
Ø 32
152
44
148
133
12
R15
286
209
143
END VIEW
304
101
101
12
66
8
117
75
Brass Desk Turn
FRONT VIEW

APPENDIX

NEW MUTATIONS

As I draw this book to a close, so news reaches me from Europe of a number of further mutations and composite forms which have been established during the past few years which are not yet known in Gt Britain. My knowledge of European languages is rather limited and it has not always been easy to establish a complete picture of all of these birds. The date and place of their emergence has not yet been ascertained but no doubt all these will fall into place in the course of time.

The Phaeo

The most attractive kind, to my mind, is a mutation called *Phaeo* which can be had in various other mutations. The main features of this form is its reddish brown colouring on head, chest and markings on wings. When combined with the Orange Brested (see page 62) it is quite striking and another form is similar to the Chestnut Flanked White, only with reddish brown markings. The Black Breasted and Orange Breasted characters seem to have been used quite a lot in the production of *Phaeo* types and very pleasing they are too.

White Breasted

A White Breasted form is also frequently mentioned on its own and in combination with other mutations. However from the description and colour plates of such birds, they would appear to be very similar if not the same as our well known Penguins. It will need a careful examination of these birds to ascertain whether or not they are new or just another name for Penguins. At one time Penguins were known in certain countries in Europe as Silverwings and Whitebellies. This certainly points to their being a well known Penguin.

Pastel

Another colour phase which is a Dilute form and called Pastel produces some very nice pale coloured specimens of a shade somewhere between Silver and White. Here again, these birds could be a further form of Silver or they could have been evolved by selective breeding of Silvers or they could be an entirely new mutation. These birds can be had in both Silver and Cream shades, with a very pallid depth of colour and often combined with other mutations. There is also a Dominant Pastel Grey which is very much like our Dominant Silver, so the Pastel could be something quite new.

Grey and Fawn

The Grey (lead) Cheeked kind (see page 62) is also described as being in grey and fawn (ear lobes brownish-grey) kinds of the Chestnut Flanked White type and also in the Normal Grey and Fawn types. As this Grey Cheeked kind can be had in the forms mentioned above it should be possible to produce them in a further range of colours.

Variation of Florida Fancy Whites

There is also a form like the Florida Fancy White (see page 60) which have a white or cream body with cheek lobes and flankings, but no tear marks or chest stripes and bar. These birds could have been derived from the Florida Fancy White stock or they could be a further mutation as it is known that a like type occurred in Australia before the Florida Fancy Whites turned up in America.

When all these types are bred in greater numbers, no doubt examples will filter through to Gt Britain and then their colour types can be analysed. As it is well known there are variations of shades within each mutation and a mutation bred in one area may appear slightly different from those bred in another. Until a careful assessment of each mutation has been made, there is certain to be differences in naming mutations within individual countries.

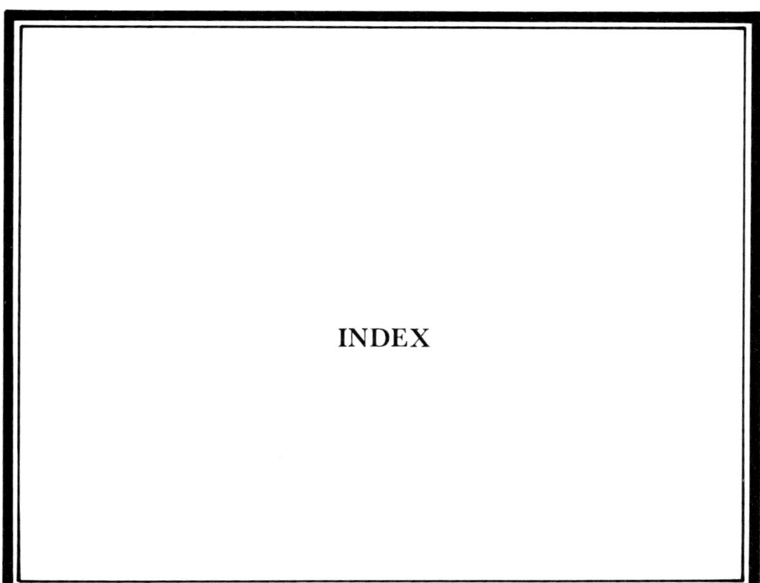

INDEX

INDEX

Page numbers in *italics* denote illustrations

I

K

L